NANOTECHNOLOGY SCIENCE AND TECHNOLOGY

DYNAMICS OF INFILTRATION OF A NANOPOROUS MEDIA WITH A NONWETTING LIQUID

NANOTECHNOLOGY SCIENCE AND TECHNOLOGY

Additional books in this series can be found on Nova's website under the Series tab.

Additional E-books in this series can be found on Nova's website under the E-books tab.

NANOTECHNOLOGY SCIENCE AND TECHNOLOGY

DYNAMICS OF INFILTRATION OF A NANOPOROUS MEDIA WITH A NONWETTING LIQUID

V. D. BORMAN
AND
V. N. TRONIN

Nova Science Publishers, Inc.
New York

Copyright © 2010 by Nova Science Publishers, Inc.

All rights reserved. No part of this book may be reproduced, stored in a retrieval system or transmitted in any form or by any means: electronic, electrostatic, magnetic, tape, mechanical photocopying, recording or otherwise without the written permission of the Publisher.

For permission to use material from this book please contact us:
Telephone 631-231-7269; Fax 631-231-8175
Web Site: http://www.novapublishers.com

NOTICE TO THE READER

The Publisher has taken reasonable care in the preparation of this book, but makes no expressed or implied warranty of any kind and assumes no responsibility for any errors or omissions. No liability is assumed for incidental or consequential damages in connection with or arising out of information contained in this book. The Publisher shall not be liable for any special, consequential, or exemplary damages resulting, in whole or in part, from the readers' use of, or reliance upon, this material.

Independent verification should be sought for any data, advice or recommendations contained in this book. In addition, no responsibility is assumed by the publisher for any injury and/or damage to persons or property arising from any methods, products, instructions, ideas or otherwise contained in this publication.

This publication is designed to provide accurate and authoritative information with regard to the subject matter covered herein. It is sold with the clear understanding that the Publisher is not engaged in rendering legal or any other professional services. If legal or any other expert assistance is required, the services of a competent person should be sought. FROM A DECLARATION OF PARTICIPANTS JOINTLY ADOPTED BY A COMMITTEE OF THE AMERICAN BAR ASSOCIATION AND A COMMITTEE OF PUBLISHERS.

LIBRARY OF CONGRESS CATALOGING-IN-PUBLICATION DATA

Borman, V. D.
 Dynamics of infiltration of a nanoporous media with a nonwetting liquid / V.D. Borman and V.N. Tronin.
   p. cm.
 Includes index.
 ISBN 978-1-61668-865-3 (softcover)
 1. Porous materials--Fluid dynamics. 2. Seepage--Mathematical models. 3. Liquid metals--Viscosity. 4. Nanostructured materials. I. Tronin, V. N. II. Title.
 TA418.9.P6B67 2009
 620.1'16--dc22
                                    2010016668

Published by Nova Science Publishers, Inc. ✦ *New York*

CONTENTS

Preface		vii
Chapter 1	Introduction	1
Chapter 2	Experimental Technique and Results	9
Chapter 3	Model of Infiltration Dynamics for a Porous Media	19
Chapter 4	Discussion of Results and Comparison with Experiment	53
Acknowledgments		61
References		63
Index		65

PREFACE

After compression of a system formed by a nanoporous media and a nonwetting liquid to the threshold pressure value, the liquid fills the pores of a porous media. In accordance with prevailing concepts, passage of the liquid from the bulk to the dispersed state can be described as a percolation-type transition. This process is typical of infiltration of macroscopic porous bodies with wetting liquids. The threshold type of infiltration was observed for nonwetting liquids and is scientifically detailed in this book.

Chapter 1

1. INTRODUCTION

After compression of a system formed by a nanoporous media and a nonwetting liquid to the threshold pressure value p_{c0}, the liquid fills the pores of a porous media. In accordance with prevailing concepts, passage of the liquid from the bulk to the dispersed state can be described as a percolation-type transition [1]. The percolation-type spatial distribution of clusters formed by pores filled with the liquid is confirmed by the "devil's staircase" effect involving the change in the resistance of a porous media (porous glass) upon its infiltration with mercury in the vicinity of the threshold infiltration pressure [2]. The percolation type of infiltration of porous bodies is also confirmed by the "viscous fingers" effect, in which a wetting liquid is displaced from pores by some other liquid [3]. In this case, a nonuniform front of porous media infiltration is formed. This process is typical of infiltration of macroscopic porous bodies with wetting liquids. The threshold type of infiltration was observed for nonwetting liquids, for grained porous bodies (zeolites) with a pore size of $R = 0.3$–1.4 nm and silochromes ($R = 4$–120 nm) filled with nonwetting liquid metals, and for hydrophobized granular porous bodies with a silicon oxide skeleton ($R = 3$–50 nm) filled with water, ethylene glycol, or salt solutions [4–20]. The grain size in [4–20] was 1–100 μm.

To fill nanometer-size pores with a nonwetting liquid with a surface energy of 0.05–0.50 J/m^2, a threshold pressure of $p_{c0} = 10^2$–10^3 atm is required. When the liquid passes from the bulk to the dispersed state in a nanoporous media with a specific volume of 1 cm^3/g, the energy absorbed by the liquid and returned (accumulated) when the liquid flows out amounts to 10–100 kJ/kg. This value is an order of magnitude higher than for polymer composites or alloys with the shape memory effect, which are widely used now [21]. This forms the basis for devices for mechanical energy absorption and accumulation. Bogomolov [22] was the first to indicate such a possibility of accumulating mechanical energy. It should be noted that 1 kg of a porous material is sufficient for absorbing the energy of a media having a mass of 1 t and moving at a velocity of 50 km/h.

In earlier publications, infiltration of pores in a porous media was described in the mean field approximation as a percolation transition in an infinitely large porous media [1]. The pore volume filled under a pressure p was calculated as a fraction of the volume of an infinitely large cluster formed by pores with a radius larger (in accordance with the Laplace pressure) than the minimal radius of the pores accessible to the nonwetting liquid under the given pressure. The mean field approximation using the Bethe lattice makes it possible to qualitatively describe the dependence of the filled volume on pressure in the vicinity of threshold p_{c0} only under the assumption of a special asymmetric size distribution for pores [1].

In contrast to second-order phase transitions including the percolation transition [2], the systems under investigation exhibit an infiltration-defiltration hysteresis, as well as (complete or partial) nonoutflow of the nonwetting liquid from the porous media when the excess pressure drops to zero [4, 5, 8, 20]. It should be noted that the nondefiltration restricts the application of the system for energy absorption and accumulation, while hysteresis controls the absorbed and accumulated energy (returned during defiltration) [23].

It was shown in [8, 20] that during slow infiltration of the systems under investigation, the pressure dependence of variation $\Delta V(p)$ in the volume of liquid in a porous media in the infiltration-defiltration cycle (hysteresis) and the volume of the liquid remaining in pores can be described by percolation theory if we take into account energy barrier $\delta A(R, p)$ of the fluctuational infiltration-defiltration of the liquid in a pore of radius R. Condition $\delta A(R, p) = 0$ for porous medias with a certain pore size distribution makes it possible to find the pressure that corresponds to the access of a pore of radius R to infiltration in a system of connected pores. For porous bodies, this condition generalizes the Laplace relation. With increasing pressure, the number of pores accessible to infiltration increases and the pores surrounding the given one may also become accessible. Thus, a cluster of accessible pores filled with liquid is formed in the porous media.

Introduction

For the systems studied in [8, 20], infiltration of porous medias upon a slow change in pressure is observed in the vicinity of the percolation threshold for such a fraction $\theta(p)$ of the volume of accessible pores, for which the inequality $[\theta_{c0} - \theta(p)]/\theta_{c0} \ll 10^{-2} - 10^{-4}$ holds, where θ_{c0} is the percolation transition threshold ($\theta_{c0} = 0.18$ for 3D systems [24, 25]). This means that when the grain size of the porous media is $L \sim (10^2\text{-}10^4)\, \overline{R}$, where $\overline{R}$ is the mean pore radius in a grain ($\overline{R} \sim$ 1-10 nm), correlation length $\xi \approx \overline{R}/|\theta - \theta_{c0}|^\nu$ ($\nu = 0.8$ [24, 25]) becomes comparable to grain size L or exceeds it ($\xi \geq L$). This allows us to treat the infiltration of a grain of the porous media as a spatially uniform process by which clusters of filled pores form.

If the characteristic time τ_p of variation in pressure is much longer than characteristic hydrodynamic time τ_z, of nonthreshold ($\delta A(R, p) = 0$) infiltration of clusters of accessible pores, the volume of the liquid in the porous media at a given pressure can be calculated if the distribution function for accessible-pore clusters over the number of pores in them is known [20]. Infiltration first occurs from the grain surface, and then the liquid flows via clusters of filled pores to the clusters of accessible pores. It should be noted that in all experiments [4-20], the measurement sensitivity of the filled volume was $\delta \Delta V / \Delta V = 5\text{-}10\%$ and the initial stage of infiltration of the surface layer of a grain was not detected. Thus, infiltration of grains of a porous media with a non-wetting liquid for $\tau_p \gg \tau_z$ can be described as infiltration of clusters formed by accessible pores. In view of the small grain volume, we can disregard the spatial non-uniformity in the formation of clusters of accessible pores.

It was shown in [26, 27] that upon rapid compression (with a pressure growth rate of $\dot{p} = 10^4\text{-}10^5$ atm/s) of the systems formed by a silochrome SKh 1.5 granulated porous media and Wood's alloy or a Fluka 100 hydrophobized granulated porous media and water, infiltration takes place beyond the percolation threshold at a pressure considerably exceeding threshold pressure p_{c0}. The threshold pressure was $p_0 = 1.6 p_{c0}$ for the former system [27] and $p_0 = 2 p_{c0}$ for the latter system [26]. Infiltration is also associated with irregular oscillations in pressure [27]. It follows hence that when the characteristic time of compression of the system decreases, the mechanism of infiltration of the porous media changes. However, the mechanism of infiltration of the porous media under strong compression remains unclear.

To reveal the regularities of infiltration of a nanoporous media with a nonwetting liquid is of fundamental importance for understanding the dynamics of percolation transition and of practical interest for the development of shock-absorbing systems.

In Section 2, we will study experimentally the infiltration-defiltration process for systems consisting of a Libersorb 23 (L23) hydrophobic granular nanoporous media and water or an aqueous solution of $CaCl_2$ for pressure compression rates of $\dot{p} > 10^4$ atm/s in the situation when the characteristic time τ_p of pressure growth is shorter than the characteristic time τ_z of non-threshold hydrodynamic infiltration of clusters of accessible pores. New regularities in threshold infiltration under rapid compression are established, which noticeably distinguish the infiltration process in this case from infiltration of a nanoporous media under slow variation in pressure.

It can be expected that upon an increase in compression rate and a decrease in time τ_p as compared to τ_z, the fraction of accessible pores increases and the system is "thrown" beyond the percolation threshold. In this case, an "infinitely large" cluster of accessible pores is formed in each grain, and the fraction of such pores increases so that the medium of pores in the grain becomes virtually homogeneous. Consequently, upon a decrease in ratio τ_p/τ_z, infiltration must be in compliance with the Darcy law [28] upon an increase in pressure, and the infiltration time of the porous media must decrease. However, it was found that infiltration pressure p_0 in the systems under investigation is independent (within the experimental error) of the compression energy and, hence, of the pressurization time. During infiltration of the porous media, the new value of threshold pressure p_0 remains unchanged and the filled volume is determined not by the fraction of accessible pores, but by the compression energy. For $p < p_0$, the liquid does not infiltrate the porous media. Thus, it was found that pulsed compression of the systems studied here leads to the emergence of a threshold infiltration pressure p_0 higher than pressure p_{c0} of the percolation transition observed for $\tau_p \gg \tau_z$. It was also found that the area of the infiltration-defiltration hysteresis loop under rapid compression is larger than

Introduction 5

for $\tau_p \gg \tau_z$. This indicates the emergence of an additional dissipation mechanism. We can naturally It associate this additional dissipation with a flow of the viscous liquid in a porous media. It was found, however, that the experimental time dependences of pressure and volume for the systems studied here do not change (within experimental error) upon a fivefold change in the viscosity of the liquid. Thus, it is established that the infiltration rate in grains of a hydrophobic nanoporous media is independent of the viscosity of the liquid.

In Section 3, a model describing the dynamics of infiltration in a granular porous material is constructed. It is assumed that infiltration in grains occurs independently and a pressure-dependent distribution of accessible pore clusters is formed in each grain. Under fast compression, infiltration occurs at a pressure of $p_0 > p_{c0}$. For the systems studied here, $p_0 \approx 1.2 p_{c0}$, and more than 70% of all pores become accessible to infiltration.

According to estimates, infiltration of a grain of a porous media under rapid compression occurs when the fraction of accessible pores is $\theta_0 = 0.28$, which is higher than percolation threshold $\theta_{c0} = 0.18$. In this case, the porous media is beyond the percolation threshold for accessible pores and an infinitely large cluster of accessible pores (whose size coincides with the size of the grain), surrounded by smaller clusters of accessible pores, is formed in each grain of the porous media. Finite-size clusters contain about 20% of all pores in the porous media, while the infinitely large cluster contains 80% of all accessible pores. For this reason, infiltration in a grain of the porous media under rapid compression will be described as rapid infiltration of liquid into finite-size clusters of accessible pores occurring simultaneously in the entire space of pores in a grain, followed by slow percolation of the liquid from these clusters into the growing infinitely large cluster of accessible pores. Obviously, no infiltration front is formed in this case over time intervals of percolation of the liquid into the infinitely large cluster.

For $\tau_p > \tau_z$, infiltration of a liquid into a porous media is described as percolation of the liquid from a cluster of filled pores to a cluster of accessible pores, while for $\tau_p < \tau_z$, the process is the percolation of the liquid from a cluster of filled pores to the infinitely large cluster of accessible pores. We solve a system of kinetic equations constructed for coordinate-independent distribution functions for clusters of accessible and filled pores and which describe these process is solved for slow and fast infiltration.

In the case of slow infiltration, a new result is the divergence of the characteristic time τ_v of infiltration in pores of a grain at percolation threshold θ_{c0} via accessible pores (critical retardation). In the case of fast infiltration, solution of the system of kinetic equation implies that infiltration must occur at a constant pressure p_0. For $p < p_0$, infiltration should not be observed. Pressure p_0 and characteristic time τ_v are controlled by the characteristic time of pressurization in the vicinity of the new value of infiltration threshold θ_c, which is higher than the known percolation threshold. Quantity θ_c is a universal characteristic for porous bodies, and pressure $p_c < p_0$ corresponding to it is determined by the size distribution of pores and by surface energies of the liquid and the interface between the liquid and the porous media.

The solution of the system of kinetic equations leads to another new result, viz., nonlinear response of the medium to external action, which is manifested in the compensation of this action due to percolation of the liquid from the cluster of filled finite-size pores to an infinitely large cluster of accessible but unfilled pores. As a result of such compensation, infiltration must be independent of the viscosity of the liquid. Infiltration must be accompanied by oscillations of pressure and smaller (in absolute magnitude) oscillations of the volume.

The resultant time dependences of pressure and volume under rapid compression, as well as the dependences of p_0, the maximum filled volume, and the total infiltration time on the compression energy, successfully describe the experimental data for systems L23 + H_2O and L23 + $CaCl_2$ under investigation (Section 4). The domain of applicability of the proposed model of infiltration dynamics of nanoporous bodies is also considered in this section.

This work is the result of rsearch carrrier out in following articles: V. D. Borman, A. M. Grekhov, and V. I. Troyan, Zh. Éksp. Teor. Fiz. 118 (1), 193 (2000) [JETP 91 (1), 170 (2000)], V. D. Borman, A. A. Belogorlov, A. M. Grekhov, G. V. Lisichkin, V. N. Tronin, and V. I. Troyan, Zh. Éksp. Teor. Fiz. 127 (2), 431 (2005) [JETP 100 (2), 385 (2005)], V. D. Borman, A. A. Belogorlov, A. M. Grekhov, G. V. Lisichkin, V. N. Tronin, and V. I. Troyan, Pis'ma Zh. Tekh. Fiz. 30 (23), 1 (2004) [Tech. Phys. Lett. 30 (12), 973

(2004)], V. D. Borman, A. A. Belogorlov, A. M. Grekhov, V. N. Tronin, and V. I. Troyan, Pis'ma Zh. Éksp. Teor. Fiz. 74 (5), 287 (2001) [JETP Lett. 74 (5), 258 (2001)], V. D. Borman, A. A. Belogorlov, G. V. Lisichkin, V. N. Tronin, and V. I. Troyan JETP, Vol. 108, No 3, March 2009.

Chapter 2

2. EXPERIMENTAL TECHNIQUE AND RESULTS

In experiments, we studied the dynamics of infiltration of water and aqueous solutions of $CaCl_2$ in Libersorb 23 (L23) granular nanoporous media with a mean pore radius of $\overline{R} \approx 6.5$ nm. This porous media is KSK-G silica gel with SiO_2 as the skeleton material, whose surface was chemically modified in accordance with the technique described in [29] to impart hydro-phobic properties to the surface. The specific surface of L23 is approximately 200 m^2/g, its specific volume is 0.56 cm^3/g, and the mean grain size of the powder of the porous media is 10 |Lim. A sample of the porous media 2-10 g in mass was placed in a container permeable to the liquid in a high-pressure chamber with a volume of ~60 cm^3. The chamber was filled with a liquid (water or 25% (in mass) aqueous solution of $CaCl_2$). A movable 180-mm-long rod 10 mm in diameter was inserted through a seal in the cover of the chamber.

In experiments on infiltration in nanoporous bodies, the liquid-porous media system was subjected to fast compression on the experimental bench shown schematically in Figure 1. Lower slab *1* is fixed by mounts *2* to upper slab *3*. Load *5* 10 kg in mass could freely slide over steel ropes *4*. Strain gauge *6* bearing high-pressure chamber *7* filled with a liquid and a porous media was fastened to slab *1*. The gauge could measure forces from 10 to 10^4 N with an error less than 5% for forces exceeding 100 N. Rod *8* of the chamber was rigidly connected with the rod of displacement pickup *10* via steel plate *9*. During the impact against load *5*, rod *8* entered chamber *7*, leading to an increase in pressure in the system. Pickup *10* detected displacements of rod *8* of up to 14 cm under impact and changes in the volume (ΔV) of up to 11 cm^3 (for area 5-0.8 cm^2 of the rod) with an error not exceeding 5%. Gauge 6 measured force F exerted by the load on the rod and, hence, the pressure in the chamber ($p = F/S$). The frequency range of the force and displacement pickups with a constant sensitivity was limited by a frequency of 5 kHz. Signals from the pickups were detected via an analog-to-digital converter and processed by a computer. The pressurization rate in experiments was $\dot{p} = (1\text{-}8) \times 10^4$ atm/s. Energy E of the impact varied from 20 to 100 J.

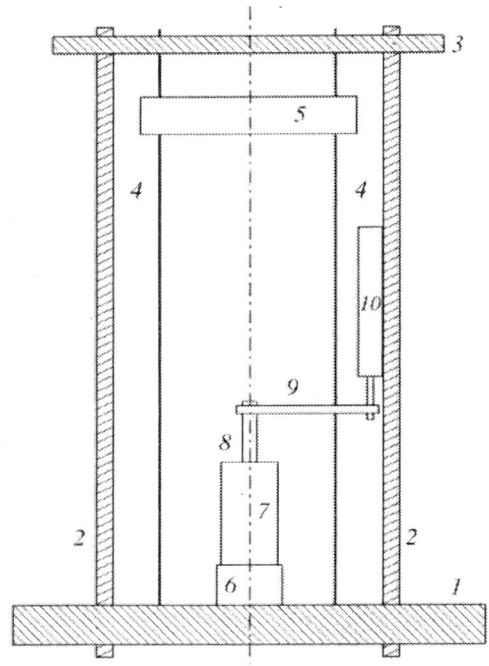

Figure 1. Experimental bench for studying the dynamics of infiltration of a nonwetting liquid in nanoporous bodies.

Infiltration of water (L23 + H$_2$O) and 25% aqueous solution of CaCl$_2$ (L23 + CaCl$_2$) into porous media L23 for a low pressurization rate ($\dot{p} \leq 1$ atm/s) was also studied for comparison. For this purpose, we used the setup described in [20], which ensured a slow variation in pressure and measurement of the change in volume of the system (i.e., the volume of the liquid infiltrating the porous media at a fixed pressure). Additionally, total compressibility of the chamber and the liquid (χ = (4.5 ± 0.4) x 10^{-3} cm^3/atm for water and χ = (3.1 ± 0.3) x 10^{-3} cm^3/atm for an aqueous solution of CaCl$_2$), as well as compressibility χ = (1.8 ± 0.2) x 10^{-3} cm^3/atm of the empty porous media, were measured in experiments when the chamber was filled with a liquid without a porous media. The reproducibility of infiltration—defiltration of the CaCl$_2$ solution in the porous media indicated the absence of segregation of the salt in the pores of L23.

Figure 2 shows the time dependences of pressure in the chamber filled only with the liquid (aqueous solution of CaCl$_2$ with a volume of 60 cm^3), as well as the time dependence of pressure and volume in the case when the chamber was filled with the aqueous solution of CaCl$_2$ (with a volume of 55 cm^3) and the L23 porous media with a mass of m = 4 g. These curves were obtained for an impact energy of E = 40 ± 2 J. For the L23 + H$_2$O and L23 + CaCl$_2$ systems studied here at pressurization rate $\dot{p}$ > 10^4 atm/s, irregular oscillations of pressure took place; similar oscillations were observed in the liquid Wood's alloy-silochrome SKh-1.5 system [27]. It should be noted that the amplitude of irregular volume oscillations predicted in [27] is much lower in the systems studied here than the amplitude of pressure oscillations. It follows from Figures 2a and 2b that in contrast to elastic compression of the chamber with the liquid, the increase in pressure in the porous media-liquid system is limited by the value of pressure p_0 = 205 ± 10 atm averaged over irregular oscillations. Figure 2b shows for comparison the value of pressure p_{c0} = 180 atm corresponding of the threshold of infiltration of the CaCl$_2$ solution into pores of L23 for a low pressurization rate $\dot{p}$ ≤1 atm/s. The threshold values of pressure for the L23 + H$_2$O system are p_{c0} = 150 ± 8 atm and p_0 = 180 ± 9 atm. Quantity p_{c0} is defined as the pressure at which the compressibility of the infiltrated liquid-porous media system is maximum. The characteristic time of the increase in pressure from p_{c0} to p_0 is t_1 = 2 ms, which corresponds to a pressurization rate of $\dot{p}$ ≈ 4x10^4 atm/s. It can be seen that for $\dot{p}$ > 10^4 atm/s, the volume of the system decreases at a pressure p_0 higher than the percolation transition pressure p_{c0} [8, 20] in the case of slow infiltration. The duration of compression of the system is controlled by time t_2 = 23 ms, at which the decrease in volume is maximum. During the time interval from zero to t_1, the decrease in the volume of the system is $-\Delta V$ = 1.10 ± 0.05

cm^3 and is equal (to within the measurement error) to the decrease in volume $-\Delta V = 1.00 \pm 0.05$ cm^3 due to compressibility of the chamber, liquid, and porous media. In time interval $t_1 - t_2$, the value p_0 of the pressure averaged over oscillations is constant; consequently, the observed change in the volume (see Figure 2c) is associated not with the compressibility of the chamber and system, but with the infiltration through the pores of the porous media. Thus, infiltration of pores begins at a pressure p_0 higher than the percolation transition pressure; maximum infiltration (change in the volume of the system) is attained at instant t_2, and the entire process of infiltration occurs at a constant pressure p_0 averaged over oscillations. Maximal infiltration at $t = t_2$ is $\Delta V_m = 1.20 \pm 0.05$ cm^3, which is smaller than the volume $V_{pore} = 2.3$ cm^3 of pores in the sample; i.e., for impact energy $E = 40$ J, infiltration of liquid through accessible pores in the sample with the mass of $m = 4$ g does not occur. According to estimates, the work of compression ($E_{el} = 42 + 2$ J) in the time interval from 0 to t_2 coincides with impact energy $E = 40 + 2$ J to within the measurement error. Over time intervals $t > t_2$, the increase in the volume of the system and chamber is associated with the removal of elastic stresses and defiltration of the liquid from the pores of the porous media. Dependences analogous to those depicted in Figure 2 are also observed for the L23 + H$_2$O system.

Our measurements make it possible to find the dependence of infiltration pressure p_0, the maximum filled volume of pores $\Delta V_m = \Delta V(t_2) - \Delta V(t_1)$, and infiltration time $t_{in} = t_2 - t_1$ under rapid compression of the system on infiltration energy $E_{in} = E - E_{el}$, where E_{el} is the part of the impact energy spent for elastic compression of the liquid-porous media system and on the increase in the volume of the chamber, $E_{el} = (\chi_1 + \chi_2 + \chi_3) \cdot p_0^2 / 2$. Since the value of p_0 is independent of the impact energy (Figure 3a), the value of E_{el} is constant to within the measurement error. Figure 3a shows that infiltration pressure p_0 for the L23 + H$_2$O system is independent of energy to within experimental error in the range $E_{in} = 30-80$ J. However, a tendency toward an increase in p_0 is observed upon an increase in energy. The dependences of the infiltration time for a porous media and maximum filled volume ΔV_m of pores on the specific energy of infiltration are close to linear to within the measurement error (Figures 3b and 3c). The $\Delta V_m(E_{in})$ curve is plotted for the specific energy of infiltration (E_{in}/m). The maximum possible filled volume is limited by the specific volume of pores and is proportional to the mass of the porous media. The possible maximum energy absorbed during infiltration is also proportional to the mass of the porous media. For L23, the specific volume of pores is 0.56 cm^3/g. Dependence $\Delta V_m(E_{in}/m)$ is limited by this volume, which corresponds to the maximum specific infiltration energy (12 J/g). Analogous dependences are also observed for the L23 + CaCl$_2$ system.

It follows from dependences $\Delta V_m(E_{in})$ and $t_{in}(E_{in})$ that total flux (flow rate) J of the liquid averaged over the infiltration time is independent of energy. Indeed, the maximum infiltration volume for infiltration energy E_{in} can be defined as

$$\Delta V_m = \int_0^{t_{in}} J(t) \cdot dt = \overline{J} \cdot t_{in}$$

For $\Delta V_m \propto E_{in}$ and $t_{in} \propto E_{in}$, we have $\overline{J}(E) = const$. The same result follows from the expression for infiltration energy:

$$E_{in} = \int_0^{t_{in}} p(t) \cdot J(t) \cdot dt$$

For $p(t) = const = p_0$, we have $E_{in} = p_0 \cdot \overline{J} \cdot t_{in}$, and average flux $\overline{J}$ either depends on energy only slightly, or is independent of energy altogether to within the measurement error. It follows from Figure 2c that time dependence $\Delta V(t)$ of the sample volume deviates from the linear dependence within the measurement error only in the vicinity of the maximum infiltration time t_2. Consequently, the flux is independent of energy ($J(t) = const$) everywhere except in this neighborhood. Thus, the pressure at which infiltration through nanopores of a disordered porous media occurs for the systems under investigation, as well as the average flux of the liquid in pores in a rapidly pressurized system (when $\dot{p} > 10^4$ atm/s), depends weakly on the impact energy or is independent of it altogether, and it is apparently controlled only by the properties of the system.

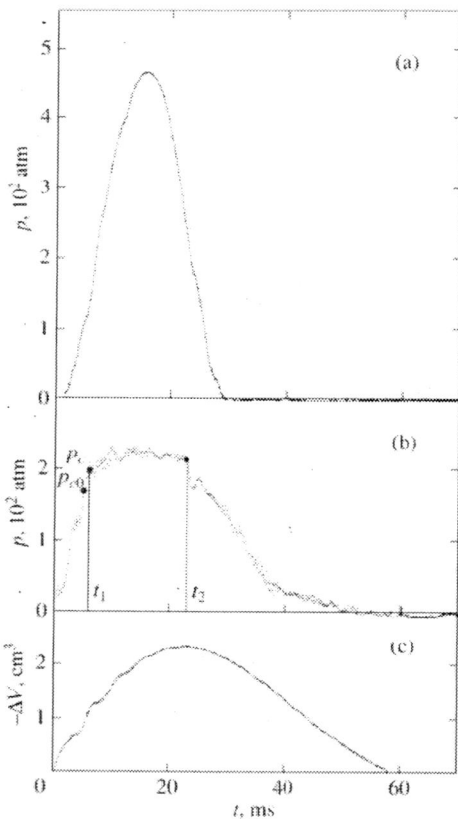

Figure 2. Time dependence of pressure in a chamber filled with a $CaCl_2$ solution (25%) (a) without a porous media and (b) with it and (c) time dependence of the decrease in the volume of the L23 + $CaCl_2$ system.

Figure 4 shows the dependence of pressure on the change in the volume of the L23 + H_2O system for various impact energies. These curves are plotted as a result of computer processing of measured dependences $p(t)$ and $\Delta V(t)$ and make it possible to analyze the features of the transformation of the mechanical impact energy during infiltration-defiltration of a non-wetting liquid in the pores of a porous media. Dependence $p(\Delta V)$ obtained for the same system for $\dot{p} < 1$ atm/s (curve 6) is also shown in Figure 4 for comparison. Under such conditions, the initial increase in pressure under elastic compression of the system and a small change in the volume in the vicinity of pressure $p_0 = 150\pm10$ atm is replaced by a decrease in the volume during infiltration of the liquid through the pores upon a small ($\Delta p/p_{c0} \approx 5\%$) change in pressure.

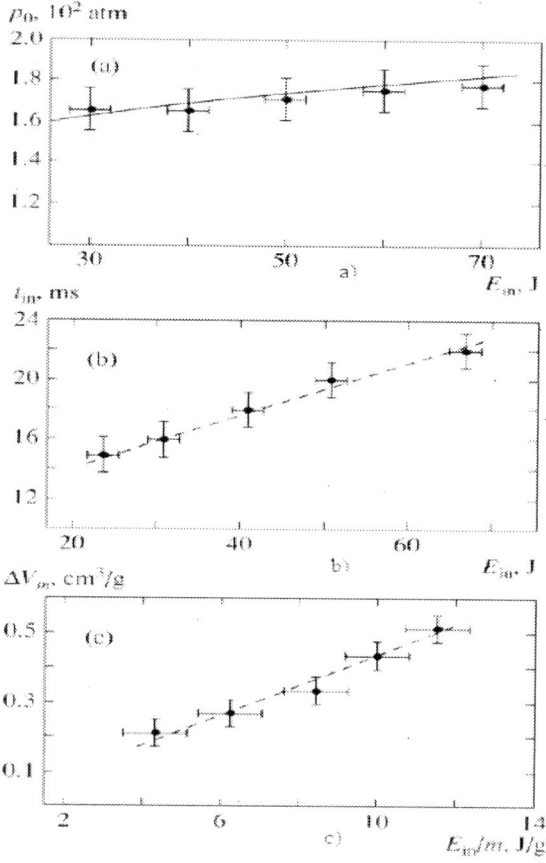

Figure 3. Dependences of (a) pressure and (b) pore infiltration time on the infiltration energy and (c) dependence of the maximum specific filled volume of pores on the specific infiltration energy for the L23 + H$_2$O system. The solid line corresponds to dependence (61), while dashed lines correspond to (55), (56) (see below).

A further increase in pressure is associated with elastic deformation of the chamber, liquid, and porous media infiltrated with the liquid. When the rod is withdrawn from the chamber, the volume of the system increases and the pressure decreases due to defiltration of the nonwetting liquid from the pores and the removal of elastic stresses. The $p(\Delta V)$ dependences in Figure 4 form hysteresis loops whose areas determine the absorbed impact energy. It can be seen that an increase in the impact energy leads to an increase in the volume of the liquid infiltrating through the pores of the porous media. It follows from Figure 4 that infiltration for different impact energies occurs at a pressure of $p_0 = 180 \pm 10$ atm. The infiltration pressure under slow compression ($\dot{p} < 1$ atm/s) increases with the filled volume of the pores. This is associated with the size distribution of pores in the porous media since upon an increase in pressure, smaller pores become accessible to the nonwetting liquid and are filled with it [8, 20]. Apparently, a certain increase in the infiltration pressure upon rapid compression (curves 1-5 in Figure 4) with the impact energy is also associated with the size distribution of pores. Analogous dependences are observed for the L23 + $CaCl_2$ system also.

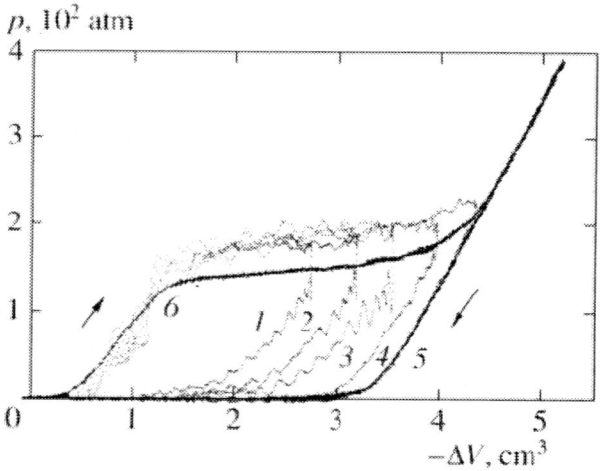

Figure 4. Infiltration–defiltration hysteresis loop for the L23 + H_2O system for various energies of impact action: $E = 30$ (1), 50 (2), 60 (3), 70 (4), and 80 J (5); curve 6 is the hysteresis loop for quasi-static infiltration–defiltration.

According to the results obtained in [20, 23], the absorbed energy is equal to the doubled energy of the formation and subsequent disappearance of menisci of the liquid in the infiltration-defiltration processes. These menisci are formed at the mouths of filled (empty) pores adjacent to the empty (filled) pores. The energy equal to the work of formation of the liquid-porous media interface and spent during infiltration is returned during defiltration of the liquid. Comparison of the $p(\Delta V)$ dependences for rapid (curves *1-5* in Figure 4) and slow (curve *6*) compression of the system shows that infiltration pressure $p_0 > p_{c0}$. It follows hence that the area of the hysteresis loop in the case of rapid compression increases and an additional dissipation mechanism apparently comes into play. The flow of the liquid in pores occurs under a considerable excess of pressure p_0 over percolation transition pressure p_{c0}. $(p_0 - p_{c0})/p_{c0} \approx 0.2$, when more than 70% of pores are accessible to infiltration of the nonwetting liquid. In this case, we can naturally assume that the additional energy dissipation is associated with energy losses in the flow of the viscous liquid in nanopores.

To verify this assumption, we performed experiments in which the $p(V)$, $p(t)$, and $V(t)$ dependences were investigated for the L23 + $CaCl_2$ system under fast and slow compression at temperatures varying from 258 to 323 K. The viscosity $\tilde{\eta}$ of the $CaCl_2$ solution varies by a factor of 5 in this temperature range [30]. However, surface energy σ of the solution [31] and surface energy $\Delta\sigma$ of the interface between the porous media and the liquid change upon variation in temperature. These quantities control pressure $p_{c0}(\sigma, \Delta\sigma)$ which determines the percolation threshold of infiltration of the solid. For 3D systems, the percolation threshold is determined by the fraction of pores accessible to infiltration, $\theta(Pco) = \theta_{co} \approx 0.18$ [3, 8, 20, 25]. Here,

$$\theta(p_{c0}) = \int_{R(p_{c0})}^{\infty} \frac{4}{3} \cdot \pi \cdot R^3 \cdot f(R) \cdot dR$$

$$R(p_{c0}) = \frac{\sigma}{p_{c0}} \cdot \left(1 + (1-\eta) \cdot \frac{\Delta\sigma}{\sigma}\right)$$

$f(R)$ is the size distribution of pores, and η is the ratio of the areas of the menisci and of the pore. This relation was obtained for an infinitely large porous media. For a real porous media with a mean pore radius $\overline{R}$ and a grain size L, the value of p_{c0} changes by approximately $\overline{R}/L = 10^{-2} - 10^{-3}$, which is smaller than the maximum experimental error.

It was found that in the temperature interval 258-323 K, the value of p_{c0} (and, hence, the values of σ and $\Delta\sigma$) do not change to within the measurement error. In the time interval from 0 to t_{in}, the values of $p(t)$, as well as of $\Delta V(t)$, obtained for different temperatures for impact energy E = 40±2 J coincide to within the experimental error (Figure 5). It follows hence that for the L23 + $CaCl_2$ system under investigation, the infiltration dynamics in the temperature range 258-323 K and the flow of liquid in nanopores are independent of the viscosity of the liquid.

Thus, for a pressurization rate of $\dot{p} > 10^4$ atm/s in the systems studied here, infiltration of nanopores of a porous media occurs at a constant pressure p_0, which is higher than percolation transition pressure p_{c0}. Pressure p_0 weakly depends on the impact energy, exhibiting a tendency toward growth within the experimental error (see Figure 3a). The energy dependence of the filled volume of pores and infiltration time are close to linear dependences to within the experimental error (see Figures 3a and 3b), and the mean flux of the liquid in pores is independent of the impact energy. During infiltration, additional dissipation energy is observed as compared to slow infiltration; however, dependences $p(t)$ and $\Delta V(t)$ do not change with temperature or upon a fivefold variation in the viscosity coefficient of the liquid (see Figure 5). It has also been established that the relative amplitude of oscillations of the volume during infiltration is considerably smaller than the relative amplitude of pressure oscillations.

Chapter 3

3. MODEL OF INFILTRATION DYNAMICS FOR A POROUS MEDIA

3.1. FORMULATION OF THE PROBLEM

Let us consider the dynamics of infiltration of grains in a disordered nanoporous media containing pores of different sizes and immersed in a nonwetting liquid. We assume that infiltration in grains occurs independently. At the initial instant, pores in each grain are empty and the liquid pressure is zero. As the pressure increases and attains a critical value, infiltration begins in the grains of the porous media. The problem involves the calculation of the time dependence of a filled volume, $V(t)$, at a preset pressure $p(t)$ with a characteristic time τ_p of increasing pressure for various relations between this time and the characteristic hydrodynamic time of infiltration of the porous media. Speaking of infiltration in the porous media, we will henceforth mean in all cases the infiltration of one of its grains, unless the opposite is specially stipulated. Obviously, infiltration may occur in a grain only if the pores form a connected system in it. Porosity φ defined as the ratio of the pore volume to the volume of a grain in the porous media must be such that the fraction of connected pores is considerably larger than the fraction of pores that do not belong to the connected system. If size L of the grains of the porous material is much smaller than the maximum size of the pores, the characteristics of a grain of the porous media are indistinguishable from the characteristics of an infinitely large media to within $R/L \sim 10^{-4}\text{-}10^{-2}$. In this case, infiltration through all pores of the gain may occur only when porosity φ exceeds percolation threshold φ_c, which is a characteristic of an infinitely large porous media. For 3D systems, the percolation threshold is $\varphi_c = 0.18$ [25]; the connectivity of pores with one another in this case is the result of the formation of infinitely large clusters of pores for $\varphi = \varphi_c$. For porous bodies with porosity φ near percolation threshold φ_c, this cluster is strongly rarefied and contains only about 1% of the total number of pores in the porous media [25, 32]. For porous bodies with a porosity $\varphi > \varphi_c$, the number of pores in the infinitely large cluster increases with φ, attaining a value of 100% for $\varphi \approx 1$.

Figure 6 shows the probability that a pore belongs to the infinitely large cluster as a function of porosity φ. It can be seen that with increasing porosity, for $\varphi \gg \varphi_c$, the value of $P(\varphi)$ tends to unity and, hence, the space of pores in a grain becomes nearly homogeneous due to the growth of the infinitely large cluster of pores. We will henceforth assume that $\varphi \gg \varphi_c$.

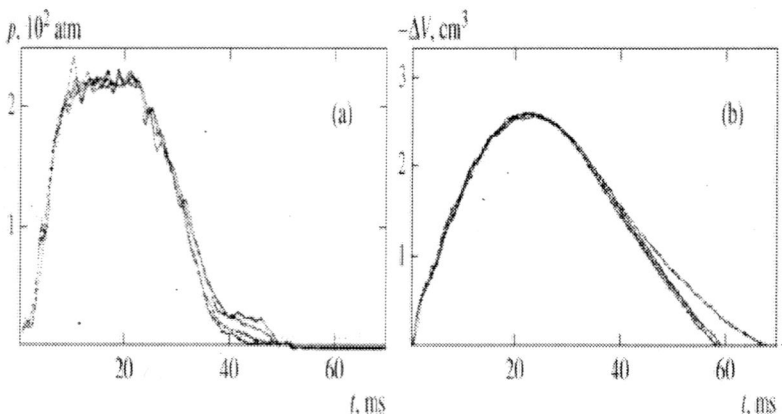

Figure 5. Time dependences of (a) pressure and (b) variation of volume for the L23 + CaCl$_2$ system with viscosity ranging from 1.27×10^{-3} to $\tilde{\eta} = 7.13 \times 10^{-3}$ Pa s. The curves in the temperature interval from 260 to 323 K coincide to within experimental error.

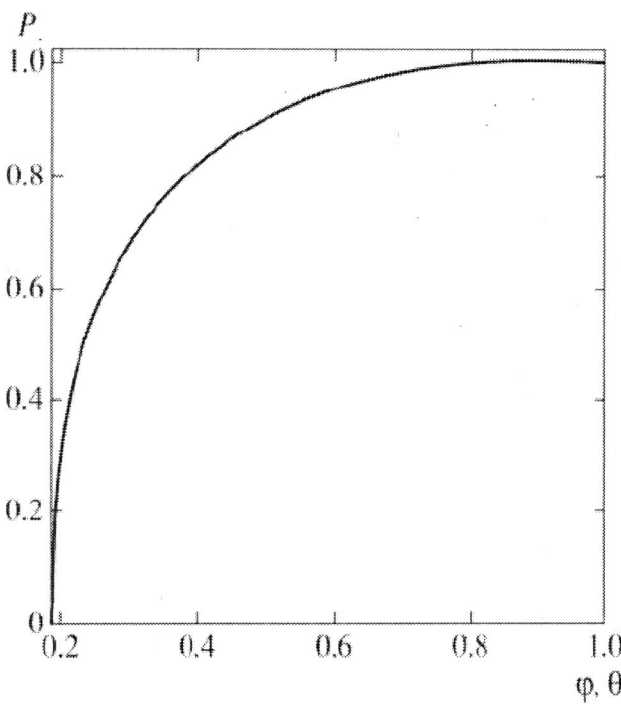

Figure 6. Probability P that a pore belongs to an infinitely large cluster as a function of porosity φ (fraction θ of accessible pores). Probability P is normalized to the total number of porosity φ or fraction θ.

The infiltration of a nonwetting liquid under a pressure p in a porous media requires that a certain amount of work be done. For this purpose, it is necessary to overcome a certain critical pressure, which is the Laplace pressure $p_c(R) \sim \delta\sigma/R$ for an isolated pore (which is assumed for simplicity to be spherical), where R is the pore radius, $\delta\sigma = \sigma^{sl} - \sigma^{sg}$, σ^{sl} and σ^{sg} being the surface energies of the interfaces between the solid and the liquid and the solid and the gas, respectively. An empty pore in a porous media may be, depending on its radius, in one of two possible states (either accessible or inaccessible to the infiltrating liquid at a given pressure p). The probability that the pore is in these states can be written in the form [20]

$$w_i(p,R) = \left[1 + \exp\left(\delta A(p,R)/T\right)\right]^{-1} \qquad (1)$$

Where

$$\delta A(p,R) = -p + \frac{3 \cdot \delta\sigma}{R} \cdot \left[1 + \eta \cdot \left(\frac{\sigma}{\delta\sigma} - 1\right)\right]$$

$\delta A(p,R)$ is the work that must be done to fill a pore of radius R with a liquid under pressure p; T is the temperature; and σ is the surface energy of the liquid.

It can be seen from expression (1) that if $\delta A(p,R) < 0$, then probability $w \sim 1$ and a pore can be filled with the liquid; if, however, $\delta A(p,R) > 0$, we have $w = 0$ and the pore becomes inaccessible. Consequently, the homogeneous space of pores with various sizes during infiltration at a preset pressure is divided into pores that can be filled, $\delta A(p,R) < 0$ (accessible pores) and pores that cannot be filled, $\delta A(p,R) > 0$ (inaccessible pores). Thus, we can assume that the medium subjected to infiltration is a heterogeneous medium consisting of accessible and inaccessible pores playing the role of white and black spheres, respectively, in percolation theory [25]. Such a medium can experience a percolation transition occurring via the formation of clusters of accessible pores followed by infiltration of a nonwetting liquid in such formations. In the general case, percolation threshold θ_{c0} for accessible pores does not coincide with φ_c. However, for $\varphi \gg \varphi_c$, in view of the homogeneity of the pore space, we can consider pores together with the skeleton material surrounding them (thick-wall pores) and analyze percolation through these pores. In this case, the percolation threshold for accessible pores and for porosity obviously coincide ($\theta_{c0} = \varphi_c$). In a porous media, pores are in contact. For this reason, the value of $p_{c0}(R)$ defined by the condition $\delta A(p,R) = 0$ is determined by the contacts of a given pore with its neighbors and, hence, on fraction η of menisci. Consequently, we can define the pores accessible at such a pressure p as pores whose radii satisfy the condition $p_{c0}(R) < p$. Upon a change in pressure, some of the formerly inaccessible pores become accessible and are filled with the liquid (if it can reach them). The approach of the liquid flow to the given pore is governed by percolation theory and occurs via the formation of accessible pore clusters both of a finite and an infinitely large size [8, 20].

Thus, the dynamics of infiltration in a grain of a porous media can be represented as the formation of the medium for infiltration (i.e., a system of clusters of accessible pores followed by infiltration in a part of these clusters). Since the infiltration in a grain of the porous media detected experimentally occurs when percolation length $\xi \approx \overline{R} / |\theta - \theta_{c0}|^{v}$ ($v = 0.8$) becomes comparable to grain size L or exceeds it $\xi \geq L$), infiltration in the grain can be treated as a uniform process occurring simultaneously in the entire pore space of the grain and resulting in the formation of clusters of filled pores.

Thus, the problem of infiltration of a porous media can be formulated as the problem of calculating the coordinate-independent distribution functions for clusters of accessible and filled pores over the number of pores, followed by calculation of volume $V(t)$ of the liquid in the porous media under pressure $p(t)$. As before [8], we assume that the size distribution for pores is narrow ($\Delta R/R < 1$) so that the percolation transition is independent of $\Delta R/R$

3.2. BASIC EQUATIONS

The times in which accessible and filled pores form are substantially different. Indeed, in accordance with expression (1), the formation of accessible pores is controlled by the time of pressure variation in the system, while the time of filling is the hydrodynamic time of infiltration of the liquid through the clusters of accessible pores. These times may differ by orders of magnitude; for this reason, the pores accessible at instant t can be divided into accessible and filled and accessible but unfilled. Consequently, to describe the infiltration dynamics, it is necessary to trace the formation processes of clusters of accessible pores and clusters of filled pores separately. In deriving kinetic equations for distribution functions $f(n,t)$ and $F(n,t)$ of accessible and filled pores, we will assume that the transformation of an accessible pore into a filled one only leads to the disappearance of the accessible pore (i.e., the infiltrated medium does not change in the course of filling). It should be noted that the change in the medium being infiltrated will be taken into account below as the filled volume is calculated in the mean field approximation.

The formation of clusters in the problem of spheres (black and white spheres) was described in [33], where the distribution function for clusters of white spheres over the number of spheres in these clusters was introduced. A change in the distribution function in this model occurs as a result of coalescence of clusters of white spheres. Following [33], we will describe the dynamics of infiltration of the liquid in a grain assuming that the medium for infiltration is inhomogeneous and consists of accessible and inaccessible pores. In this case, accessible pores play the role of white spheres and their fraction is defined as

$$\theta(p) = \int_0^\infty w(R, p) dR f_r(R) R^3 \qquad (2)$$

where $f_r(R)$ is the size distribution function for pores and quantity of $w(R, p)$ is defined by relation (1).

In describing the dynamics of infiltration of a non-wetting liquid in a porous media, pressure is a function of time; consequently, θ also depends on time. Bearing this in mind, we can write the system of kinetic equations defining the time evolution of the distribution functions for clusters of accessible and filled pores over the number of pores in them in the form

$$\frac{\partial F(n,t)}{\partial t} = \sum_{m=1}^{n-1} F(m,t) \frac{f(n-m,t)}{\tau(m,n-m)} - \sum_{m=1}^{\infty} F(n,t) \frac{f(m,t)}{\tau(n,m)} - F(n,t) \frac{S(\varepsilon(t))}{\tau_{pc}(n)} \qquad (3)$$

$$\frac{\partial f(n,t)}{\partial t} = \frac{1}{\tau_d} \{ \frac{1}{2} \sum_{m=1}^{n-1} m^q (n-m)^q f(m,t) f(n-m,t) - n^q f(n,t) \sum_{m=1}^{\infty} m^q f(m,t) - 2n^q f(n,t) S(\varepsilon) \} - \qquad (4)$$
$$- \sum_{m=1}^{n-1} F(m,t) \frac{f(n-m,t)}{\tau(m,n-m)} + \sum_{m=1}^{\infty} F(n,t) \frac{f(m,t)}{\tau(n,m)} + F(n,t) \frac{S(\varepsilon(t))}{\tau_{pc}(n)}$$

Where

$$S(\varepsilon) = \varepsilon^\delta \Theta(\theta - \theta_c),$$

$$\varepsilon = |\theta - \theta_{c0}|, \ \tau_d = \left(\frac{\partial \varepsilon}{\partial t}\right)^{-1} = (\varepsilon(t))^{1+\gamma} \tau_p, \ \tau_p = \left(\frac{dp}{pdt}\right)^{-1}; \quad (5)$$

τ_p is the characteristic time of pressure variation; τ_{pc} is the characteristic percolation time for an infinitely large cluster of accessible pores from filled clusters, τ_d has the meaning of the characteristic time in which accessible pores form upon time variation of pressure; q, δ, and γ are critical indices ($q = 0.8$, $\delta = 0.2$ [33] and $\gamma = 0.6$ for 3D systems [8]); $S(\varepsilon(t))$ is the effective part of an infinitely large cluster of accessible pores (i.e., the fraction of pores belonging to the infinitely large cluster and accessible to infiltration); and $\Theta(x)$ is the Heaviside function.

Equation (3) defines the distribution function for clusters of filled pores at an arbitrary instant. The first term describes the formation process of a cluster of n pores as a result of infiltration into clusters of $n - m$ accessible pores via clusters of m filled pores over characteristic time $\tau(m, n-m)$. The second term corresponds to the attachment of any cluster of accessible pores to the cluster of n filled pores during infiltration over characteristic time $\tau(n,m)$. The third term describes infiltration of the infinitely large cluster of accessible pores from filled clusters over characteristic time $\tau_{pc}(n)$. Equation (3) disregards the variation of distribution function $F(n,t)$ due to coalescence of clusters of filled pores with one another, which corresponds to the assumption of invariability of the medium in the course of infiltration. Function $F(n,t)$ for a nearly complete infiltration will be calculated below in the mean field approximation.

Equation (4) defines the time evolution of the distribution function for accessible-pore clusters due to their coalescence with one another (first two terms), attachment to the infinitely large cluster (third term), and infiltration-defiltration of the liquid from these clusters (three last term). Times $\tau(n,m)$ and $\tau_{pc}(n)$ appearing in Eqs. (3) and (4) can be estimated from the following considerations. Let $V(m)$ be the volume of a cluster of m accessible pores, $V(n)$ be the volume of a cluster of n filled pores, $j(n)$ be the flux from n filled pores, $S(n, m)$ be the area of contact between clusters of m

accessible and n filled pores, and $\overline{S}(n)$ be the area of the contact of the cluster of n pores with the infinitely large cluster. Then we can write $\tau(n,m) = \dfrac{V(m)}{j(n)S(n,m)}$, $\tau_{pc}(n) = \dfrac{V(n)}{j(n)\overline{S}(n)}$ hese quantities depend on the size distribution of pores. Since we are interested only in the dependences of times $\tau(n,m)$ and $\tau_{pc}(n)$ on the number of filled and accessible pores in the clusters, we will estimate the values of these quantities assuming that all pores in a cluster are of the same size coinciding with the average size of a pore in the porous media ($\overline{R}$). In this case, we have $V(m) = \dfrac{4\pi}{3}\overline{R}^3 m$, $S(n,m) = 4\pi \overline{R}^2 (nm)^q$, $\overline{S}(n) = 4\pi \overline{R}^2 n^{q'}$ (q' is the critical index). Using the known expression for the flux in a porous medium, $j = k_n \Delta p / \tilde{\eta} L$ (k_n is the penetration factor of the medium) [28], we obtain

$$\tau(n,m) = \tau_0(p) n^{-q} m^{1-q}$$
$$\tau_{pc}(n) = \tau_0(p) n^{-q'+1}$$
(6)

Where

$$\tau_0(p) = \dfrac{4\eta \overline{R} L}{3 k_n (p - p_{c0}(\overline{R}))}$$

and pressure $p_{c0}(\overline{R}) \sim \delta\sigma / \overline{R}$ is defined by the condition $\delta A(p_c(\overline{R}), \overline{R}) = 0$

Equations (3) and (4) allow us to calculate the distribution functions for clusters of accessible and filled pores over the number of pores in them for a preset variation of pressure $p(t)$. Equation (4) contains the terms with essentially different physical meaning. The first three terms in kinetic equation (4) cannot be interpreted as a collision integral since these terms vary only with $\varepsilon = \varepsilon(t)$ and $p(t)$. These terms are on the order of τ_d proportional to τ_p, which is not the intrinsic time of the system, and reflect the variation of distribution function $f(n,t)$ of accessible pores only upon the variation of pressure and, as a consequence, of quantity $\varepsilon = \varepsilon(t)$. If $\varepsilon = const$, these terms are equal to zero. For $\varepsilon = \varepsilon(t)$, these terms must appear in Eq. (4) simultaneously with $(\partial f/\partial \varepsilon)(d\varepsilon/dt)$. Thus, derivative $\partial f/\partial t$ on the left-hand side of Eq. (4), as well as derivative $\partial F/\partial t$, defines the variation of distribution functions $f(n,\varepsilon(t),t)$ and $F(n,\varepsilon(t),t)$ due to the change in the external pressure and due to infiltration–defiltration of the liquid through accessible pores.

Equations (3) and (4) contain an integral of motion corresponding to the conservation of the total number of pores accessible to infiltration taking into account the fact that part of these pores have already been filled. Indeed, multiplying Eqs. (3) and (4) by n, summing over n, and adding the resultant expression, we obtain

$$\frac{\partial}{\partial t}(\sum_{n=1}^{\infty} nF(n,t) + \sum_{n=1}^{\infty} nf(n,t)) = -\frac{d\varepsilon}{dt}\sum_{n=1}^{\infty} n^{q+1} f(n,t)S(\varepsilon) \quad (7)$$

We can write probability $P(\varepsilon)$ that an accessible pore belongs to the infinitely large cluster as

$$\frac{\partial P(\varepsilon)}{\partial \varepsilon} = \theta_d \sum_{n=1}^{\infty} n^{q+1} f(n,\varepsilon)S(\varepsilon) \quad (8)$$

where θ_d is the fraction of accessible by unfilled pores. Relation (8) is analogous to the expression derived in [33] for the problem of spheres in percolation theory. Considering that the distribution functions for clusters of accessible and filled pores depend on time both explicitly and due to the change in pressure (and, hence, in quantity $\theta(t)$, using expression (8), and setting $\theta(0) = 0$, we obtain

$$\sum_{n=1}^{\infty} nF(n,t) + \sum_{n=1}^{\infty} nf(n,t) = \theta(p(t)); \quad (9)$$

This relation corresponds to conservation of the total number of pores accessible to infiltration under pressure p at instant t. In deriving Eq. (9), we used the normalization of function $f(n,t)$ taking into account the fact that some of accessible pores may belong to an infinitely large cluster,

$$\sum_{n=1}^{\infty} nf(n,t) = \theta_d(1-P(\varepsilon))$$

In this case, distribution function $F(n,t)$ for clusters of filled pores is normalized to the total number of filled pores (including the filled pores formed from the infinitely large cluster of accessible pores).

Equations (3), (4), and (9) contain the times corresponding to different processes occurring during infiltration of a porous media: characteristic time τ_p of variation of external pressure, characteristic time τ_d of the formation of accessible pores, characteristic time $\tau_z \sim \langle \tau(n,m) \rangle$ of the formation of the cluster of filled pores (angle brackets denote averaging over the ensemble of clusters of accessible and filled pores), characteristic time $\tau_\infty \sim \langle \tau_\infty(n) \rangle$ of defiltration of the liquid to the infinitely large cluster of accessible and empty pores, and characteristic time $\tau_v \sim (\partial \sum_{n=}^{\infty} nF(n,t)/\partial t)^{-1}$ of variation of the total filled volume. For 3D systems, $\theta_{c0} = 0.18$ and $\gamma \sim 0.6$; consequently, $\tau_p > \tau_d$ in all cases in accordance with relation (5). Since infiltration of the volume occurs due to variation of the external pressure, we have $\tau_v > \max(\tau_d, \tau_z)$.

We will consider two cases corresponding to slow ($\tau_p > \tau_v \gg \tau_z > \tau_d$) and fast ($\tau_v > \tau_z > \tau_p > \tau_d$) variations of pressure. The solutions to systems of equations (3), (4), and (9) are significantly different in these cases.

3.3. KINETICS OF INFILTRATION FOR SLOW VARIATION OF PRESSURE

Let us consider the case of a slow variation in pressure, when $\tau_p > \tau_v \gg \tau_z > \tau_d$. We will be interested in infiltration of a porous media over time intervals $t \sim \tau_v$ and will calculate the time dependence of the filled volume under pressure p. In Eq. (4), the first term on the right-hand side plays the leading role since it is on the order of τ_d^{-1}, while the second term is on the order of $\tau_z^{-1} \ll \tau_d^{-1}$. Since $\tau_p \gg \tau_z$, a change in pressure rapidly leads to the formation of accessible pores (over time intervals $t > x_d$) followed by infiltration of the liquid (over time intervals $t \geq \tau_z$). In accordance with relation (9), the fraction of accessible pores decreases upon infiltration. An increase in pressure leads to the formation of and filling of pores that have become accessible. In view of condition $\tau_p \gg \tau_z$, infiltration of the solid media upon a slow variation in pressure occurs near the percolation threshold over accessible pores, remaining below this threshold. For this reason, $S(\varepsilon) = 0$ in Eq. (4), and the terms containing distribution function $F(n,t)$ for filled pores (and, hence quantity $\tau_z \gg \tau_p$) are small as compared to the terms containing τ_d and can be discarded. In this case, Eq. (4) assumes the form of the equation used in [33]; the solution to this equation is known:

$$f_0(n,t) = \frac{C(t)\Omega_n(t)}{Z(t)},$$

$$\Omega_n(t) = n^{-\tau} \exp(-r\varepsilon(t)^{1/a} n), \qquad (10)$$

$$Z(t) = \sum_{n=1}^{\infty} n\Omega_n(t)$$

Here, function $C(t)$ is controlled by the normalization of distribution $f_0(n,t)$, varies over time intervals $t \sim \tau_v$ and determines the filled volume. The critical indices for 3D systems are given by [24, 25, 33].

$\tau \approx 2.2$, $a \approx 0.9$,

$$r \approx \frac{1}{2} \int_0^{1/2} u^{-q}(1-u)^{-q} du$$

Distribution function $F_0(n, \varepsilon(t))$ for $\theta \leq \theta_{c0}$ is defined over time intervals $t \sim \tau_v, \tau_p > \tau_v \gg \tau_z > \tau_d$ by the steady-state solution to Eq. (3) in the absence of an infinitely large cluster of accessible pores, $S(\varepsilon) = 0$:

$$[\sum_{m=1}^{n-1} F_0(m,\varepsilon(t))m^q(n-m)^{q-1} f_0(n-m,t) - F_0(n,\varepsilon(t))n^q \sum_{m=1}^{\infty} m^{q-1} f_0(m,\varepsilon(t))] = \frac{\tau_z}{\tau_d} \frac{\partial F_0}{\partial \varepsilon} \quad (11)$$

In the continual limit, after the Laplace transformation in time, Eq. (3) is transformed into the Volterra equation of the second kind, which can be solved using standard methods [34]. However, we can predict the solution to Eq. (3) obtained in this way if we note that distribution function $f_0(n,t)$ is also a solution to Eq. (11) for $F_0(n, \varepsilon(t))$. Thus, for a slow variation in pressure, the distribution function for filled pores is proportional to distribution function (10) for accessible pores:

$$F_0(n, \varepsilon(t)) = C_1(t)\tilde{F}_0, \quad \tilde{F}_0 = \frac{\Omega_n(t)}{Z(t)} \quad (12)$$

where $\Omega_n(t)$ and $Z(t)$ are defined in (10). Function $C_1(t)$ varies over time intervals $t \sim \tau_v$ and controls the variation of the filled volume. It should be noted that distribution function (12) for filled pores was used by us earlier to describe experiments on infiltration of a nonwetting liquid through a porous media upon a slow variation in pressure [20].

We will derive the time dependence of the fraction of filled pores in the given case using relation (9). Substituting relations (10) and (12) into (9), we obtain

$$C(t) + C_1(t) = \theta(p(t)) \tag{13}$$

On the other hand, substituting relations (10) and (12) into Eq. (4) and considering that $\tau_v \gg \tau_z > \tau_d$, we obtain

$$\frac{dC}{dt} = -\frac{C(t)C_1(t)}{\tau_v} \tag{14}$$

Where

$$\frac{1}{\tau_v} = \sum_{n=1}^{\infty} nF_0(n)(\sum_{m=1}^{n-1} mF_0(m) \frac{f_0(n-m)}{\tau(m,n-m)} - \sum_{m=1}^{\infty} F_0(n) \frac{f_0(m)}{\tau(n,m)}) \tag{15}$$

Figure 7 shows the dependence of τ_v/τ_z on θ in the vicinity of percolation threshold θ_{c0}, which was calculated using relation (15) for $q = 0.83$ and $a = 1$. This dependence is successfully approximated by the expression

$$\frac{\tau_v}{\tau_z} \approx \frac{1}{(1-\theta/\theta_{c0})^\rho}$$

where $\rho \approx 0.4$. Thus, the value of τ_v/τ_z is always greater than unity and $\tau_v/\tau_z \to \infty$ near the percolation threshold.

Using relation (13) and taking into account that the change in the volume occurs over time intervals $\tau_v < \tau_p$, we derive from relation (14) the following equation describing the time dependence of the fraction of filled pores (and, hence, the filled volume):

$$\frac{dC_1}{dt} = \frac{C_1(t)(\theta - C_1(t))}{\tau_v} \tag{16}$$

Taking into account relation (13), we can use the solution to this equation with the initial condition $C_1(t) = \Omega_0$ to calculate the following dependences:

$$C_1(t) = \frac{\theta}{1 + \dfrac{\theta - \Omega_0}{\Omega_0} \exp(-\dfrac{\theta t}{\tau_v})}$$

$$C(t) = \frac{\theta}{1 + \dfrac{\Omega_0}{\theta - \Omega_0} \exp(\dfrac{\theta t}{\tau_v})}$$

(17)

It follows from relations (16) and (17) that an increase in pressure during slow infiltration leads to an increase in the filled volume with characteristic time $\tau_v \gg \tau_z$, which is accompanied by a simultaneous decrease in the fraction of accessible, but unfilled pores.

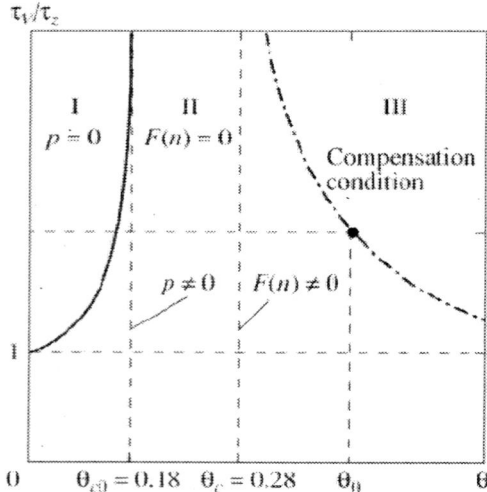

Figure 7. Dependence of relative infiltration time τ_v / τ_z on fraction θ of accessible pores for slow (I) and fast (III) infiltration (infiltration in domain II is impossible). The solid and dot-and-dash curves correspond to dependences (15) and (38), respectively.

3.4. KINETICS OF INFILTRATION UNDER A FAST VARIATION OF PRESSURE

Let us consider fast variation in pressure, when $\tau_v > \tau_z > \tau_p > \tau_d$. In this case, we will be interested, as before, in the behavior of the infiltrated porous media over time intervals $t \sim \tau_v$. As in the case of slow infiltration, the first term on the right-hand side of Eq. (4) plays the major role since it is on the order of τ_d^{-1}, while the second term is on the order of $\tau_z^{-1} \ll \tau_d^{-1}$. Since $\tau_p < \tau_z$, a change in pressure leads to rapid formation of accessible pores (over time intervals $t \sim \tau_d$). For times $\tau_z > t$, filled pores are absent. Thus, over time intervals $\tau_z > t \geq \tau_p > \tau_d$, system of equations (3), (4) assumes the form

$$\frac{\partial f}{\partial \varepsilon} = \frac{1}{2}\sum_{m=1}^{n-1} m^q (n-m)^q f(m,\varepsilon)f(n-m,\varepsilon) - n^q f(n,\varepsilon)\sum_{m=1}^{\infty} m^q f(m,\varepsilon) - 2n^q f(n,\varepsilon)S(\varepsilon); \quad (18)$$
$$F(n,t) \approx 0$$

It can be seen that over time intervals t satisfying the inequality $\tau_v > \tau_z > t \geq \tau_p > \tau_d$, the formed accessible pores now have time to be infiltrated; as a result, the porous media is in a state above the percolation threshold over accessible pores for $\theta > \theta_{c0}$ with $F(n) \ll f(n)$. Over time intervals $\tau_v > t \geq \tau_z > \tau_p > \tau_d$, the process of infiltration of the porous media begins in accordance with Eqs. (3) and (4) (in these equations, the effective part of the infinitely large cluster of accessible pores is $S(\varepsilon) \neq 0$. Over these time intervals, in view of condition $t > \tau_z \gg \tau_d$, the time derivative in Eq. (4) can be set at $(d\varepsilon/dt)(\partial f/\partial \varepsilon)$. By virtue of condition $\tau_z \gg \tau_d$, the sums of the terms in Eq. (4) containing $F(n,t)$ is zero in the zeroth and first orders in τ_d/τ_z. In this case, Eq. (3) is satisfied automatically. Thus, intervals t such that $\tau_v > t > \tau_z > \tau_p \gg \tau_d$, the equation for distribution function $f(n,t)$ for accessible pores coincides with the first equation of system (18), while the equation for $F(n,t)$ assumes the form

$$[\sum_{m=1}^{n-1} F(m)m^q(n-m)^{q-1}f(n-m,\varepsilon) - F(n)n^{q-\frac{1}{d}}\sum_{m=1}^{\infty} m^{q-1}f(m,\varepsilon)\ ] - F(n)n^{q-1}S(\varepsilon) = 0 \quad (19)$$

The equation for $f(n,t)$ at $S(\varepsilon) \neq 0$ near θ_{c0} ($\theta \geq \theta_{c0}$) has a solution differing from $f_0(n,t)$ in Eq. (10) only in the value of critical index a [33]. Function $C(t)$ appearing in Eq. (10) controls the variation of the filled volume and varies over time intervals $t \sim \tau_v$; consequently, this function can be assumed to be constant for $\tau_v > t > \tau_z \gg \tau_d$.

Equation (19) with known distribution function (10) for accessible pores is a homogeneous equation for function $F(n)$. A nonzero solution to this equation exists only if the determinant of matrix A_{nm} vanishes:

$\det(A_{nm}) = 0$

$$A_{nm} = \Delta_{nm}(n-m)^{q-1}f_0(n-m,\varepsilon)m^q - \delta_{nm}(m^q\sum_{k=1}^{\infty} k^{q-1}f_0(k,\varepsilon) + m^{q-1}S(\varepsilon)) \quad (20)$$

$$\Delta_{nm} = \begin{cases} 1, & n > m \\ 0, & n < m \end{cases}$$

Matrix A_{nm} has the form of a triangular matrix with zeros above the principal diagonal. The determinant of such a matrix is equal to the product of the diagonal elements,

$$\det A = \prod_m (-1)^m (m^q \sum_{k=1}^{\infty} k^{q-1}f_0(k,\varepsilon) + m^{q-1}S(\varepsilon))$$

and does not vanish. Consequently, Eq. (20) has no solutions for finite values of n and m. For $n \to \infty$ and $m \to \infty$, the contact areas of two clusters are controlled by a single critical index; consequently, $q \approx q' - 1$. Passing in Eq. (20) from summation to integration, considering that $f_0(n-m)\big|_{n \sim m} \propto (n-m)^{-\tau}$, and setting

$$\lim_{k \to 0} k^{q-1} f_0(k, \varepsilon) \approx 2\delta(k) \int_0^\infty dx\, x^{q-1} f_0(x, \varepsilon)$$

we obtain from Eq. (20)

$$\lim_{n,m \to \infty} A_{nm} = \lim_{n,m \to \infty} \delta_{nm} m^q (2\int_0^\infty dx\, x^{q-1} f_0(x,\varepsilon) - \int_1^\infty dx\, x^{q-1} f_0(x,\varepsilon) - S(\varepsilon)) \quad (21)$$

where δ_{nm} is the Kronecker delta. This leads to the following equation that defines the value of θ_c corresponding to a nonzero distribution function for filled pores:

$$2\int_0^\infty dx\, x^{q-1} f_0(x,\varepsilon) - \int_1^\infty dx\, x^{q-1} f_0(x,\varepsilon) - S(\varepsilon) = 0 \quad (22)$$

Expressions (2), (10), and (22) for $S(\varepsilon) \neq 0$ show that the value of θ_c is controlled by percolation threshold θ_{c0} and the critical indices appearing in Eq. (22). If function $f_0(x, \varepsilon)$ is defined by expressions (10), the integrals appearing in Eq. (22) can be expressed in terms of the gamma function and the Whittaker functions [34]. In this case, solving numerically Eq. (22) for $q = 0.83, a = 0.9$ [33], and $\theta_{c0} = 0.18$, we obtain $\theta_c = 0.28$.

Thus, Eq. (19) has the solution $F(n) = 0$ for $\theta_{c0} < \theta < \theta_c$ and $F(n) \neq 0$ for $\theta > \theta_c$. Consequently, we can state that a new state of the system is formed for $\theta > \theta_c$ over time intervals $\tau_z > \tau_p > \tau_d$ for $t > \tau_p > \tau_d$. Further infiltration of the porous media over time intervals $t \sim \tau_v$ may occur via its passage to this state, which emerges in the case under investigation due to the infinitely large cluster of accessible pores. It follows from Eq. (3) that pressure p_c corresponding to the transition of the porous media to the new state is constant and can be determined, in accordance with formula (2), from the relation

$$\int_0^\infty w(R, p_c) dRf_r(R) R^3 = \theta_c \qquad (23)$$

It follows from relations (1) and (23) that pressure p_c, in contrast to θ_c, depends on the characteristics of the porous media and the liquid, such as the size distribution function for pores, the surface energies of the liquid and porous media, and coefficient $\eta |$. By way of example, Figure 8 shows the dependence of p_c / p_{c0} on half-width $\Delta R / \overline{R}$ of the size distribution of pores at $\delta\sigma/\sigma = 1/3$, $\eta = 1.2$. Pressure p_{c0} in this case can be found from the relation

$$\int_0^\infty w(R, p_{c0}) dRf_r(R) R^3 = \theta_{c0}$$

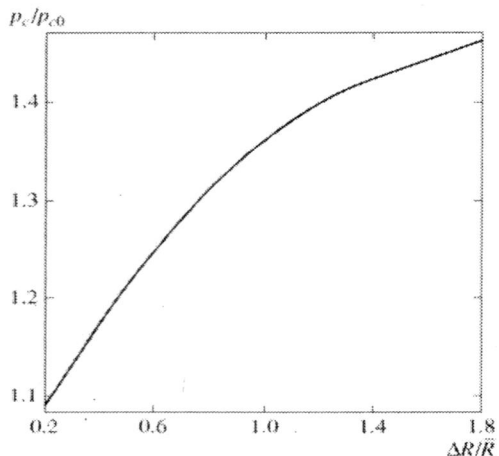

Figure 8. Dependence of ratio p_c / p_{c0} of threshold pressures on halfwidth $\Delta R / \overline{R}$ of the size distribution of pores.

Figure 8 shows that pressure p_c corresponding to the transition of the porous media to the new state is higher than p_{c0} and increases with the halfwidth of the size distribution of pores; $p_c \to p_{c0}$ for $\Delta R/\overline{R} \to 0$.

Let us now derive the equation describing the time dependence of the fraction of volume of pores filled with the liquid in the case of infiltration of a porous media in the vicinity of θ_c. For this purpose, we write Eq. (4) in the form

$$\frac{\partial F(n,\varepsilon,t)}{\partial t} = \frac{1}{\tau_0(p)} \sum_{m=1}^{\infty} A_{nm}(\varepsilon) F(m,\varepsilon,t) \qquad (24)$$

Matrix A_{nm} is defined by relation (20) and its eigenvalues are defined by the equation $\det(A_{nm} - \lambda \delta_{nm}) = 0$. For finite values of n and m, we have

$$\det(A_{nm} - \lambda \delta_{nm}) = \prod_{m}(-\lambda - (m^q \sum_{k=1}^{\infty} k^{q-1} f_0(k,\varepsilon) + m^{q'-1} S(\varepsilon))) \qquad (25)$$

and, hence, the eigenvalues of matrix A_{nm} are negative for finite n and m. For $n \to \infty$ and $m \to \infty$, we obtain, in accordance with relation (21)

$$\lambda = \lambda_\infty(\theta) \approx \langle m^q \rangle (2\int_0^\infty dx x^{q-1} f_0(x,\varepsilon) - \int_1^\infty dx x^{q-1} f_0(x,\varepsilon) - S(\varepsilon)) = z(\theta - \theta_c)^s \qquad (26)$$

Angle brackets indicate averaging over an ensemble of clusters for $m \gg 1$, and z and ξ are constants. Numerical calculations for $q = 0.83, a = 1$, and $\delta = 0.2$ give $z \approx 0.8$ and $\xi \approx 0.8$. Thus, the spectrum of eigenvalues of matrix A_{nm} for $n \to \infty$ and $m \to \infty$ acquires a small (in the vicinity of $\theta \geq \theta_c$) positive eigenvalue corresponding to relaxation time $\tau_\infty \sim (\theta - \theta_c)^{-s} \tau_z$ while the remaining eigenvalues are finite for $\theta = \theta_c$, negative, and are on the order of τ_z^{-1}.

Using relation (26), we can write Eq. (24) in the form

$$\frac{\partial F(n)}{\partial t} = \frac{\lambda_\infty(\theta)}{\tau_0(p)} F(n) + \sum_m \tilde{A}_{nm} F(m) \qquad (27)$$

Matrix

$$\tilde{A}_{nm} = \frac{1}{\tau_0(p)} A_{nm} - \frac{\lambda_\infty(\theta)}{\tau_0(p)} \delta_{nm}$$

has eigenvalues $\lambda(n) < 0$, $|\lambda(n)| \sim \frac{1}{\tau_z}$, which do not vanish for $\theta = \theta_c$. Considering that $\frac{\partial}{\partial t} = \frac{\partial}{\partial t} + \frac{d\varepsilon}{dt} \frac{\partial}{\partial \varepsilon}$, we obtain the equation

$$\frac{\partial F(n,\varepsilon)}{\partial t} + \frac{d\varepsilon}{dt} \frac{\partial F(n,\varepsilon)}{\partial \varepsilon} = \frac{\lambda_\infty(\theta)}{\tau_0(p)} F(n) + \sum_m \tilde{A}_{nm} F(m,\varepsilon) \qquad (28)$$

containing terms varying over substantially different time intervals:

$$\frac{\partial F}{\partial t} \sim \frac{F}{\tau_z}, \quad \sum_m \tilde{A}_{nm} F(m,\varepsilon) \sim \lambda(n) F \sim \frac{F}{\tau_z}$$

For time intervals $t \sim \tau_v$, we can obtain the following estimate:

$$\dot{F}(n,t) \sim \frac{F}{\tau_\varepsilon} \sim \frac{F}{\varepsilon \tau_c}, \tau_c = \frac{\tau_p}{p_c (\frac{\partial \varepsilon}{\partial p})_{p=p_c}}, \qquad (29)$$

Compressibility $(\frac{\partial \varepsilon}{\partial p})_{p=pc}$ is calculated for pressure p_c defined by relation (23), which shows that $\varepsilon(\theta_c) = \theta_c - \theta_{c0} \sim \frac{1}{x^3}, x = \frac{p_c}{p_{c0}} \geq 1$.

Consequently, $p_c(\frac{\partial \varepsilon}{\partial p})_{p=pc} \sim \frac{1}{x^5}$ and, hence, $p_c(\frac{\partial \varepsilon}{\partial p})_{p=pc} \sim (\varepsilon(\theta_c))^{\frac{5}{3}}$.

Since the value of θ_c for which the new state of the porous media being infiltrated is formed is higher than θ_{c0}, we have $p_c > p_{c0}$ and $p_c(\frac{\partial \varepsilon}{\partial p})_{p=pc} \sim 10^{-2}$. Therefore, $\tau_c \sim \tau_p \varepsilon^{-\frac{5}{3}} >> \tau_p$, for $\varepsilon << 1$. Using these estimates in the zeroth and first orders in $\frac{\tau_z}{\tau_c}$ and Eq. (28), we obtain

$$\frac{d\varepsilon}{dt}\frac{\partial F(n,\varepsilon)}{\partial \varepsilon} = \frac{\lambda_\infty(\theta)}{\tau_0(p)} F(n,\varepsilon) \qquad (30)$$

$$\frac{\partial F(n,\varepsilon)}{\partial t} = \sum_m \tilde{A}_{nm} F(m,\varepsilon) \qquad (31)$$

In fact, the procedure described above corresponds to obtaining a solution to Eq. (27) by expanding it in the eigenfunctions of operator A. Equations (30) and (31) describe substantially different processes. Equation (31) describes the kinetics of formation of finite-size clusters of filled pores over time interval $\tau_z << \tau_0 \lambda_\infty^{-1}$ around an infinitely large cluster, while Eq. (30) describes a slow "macroscopic" infiltration of the liquid into an infinitely large clusters of accessible pores through finite-size clusters of filled pores over time intervals $\tau_v \sim \tau_0 (\theta - \theta_c)^{-\xi} >> \tau_z$ for $\theta \sim \theta_c$ ($\xi \sim 0.8$). The left-hand side of Eq. (30) describes the variation in the distribution function for filled pores as a result of external action, while the right-hand side describes the variation in distribution function $F(n,\varepsilon)$ as a result of infiltration through an infinitely large cluster of accessible pores. Consequently, Eq. (30) also corresponds to the condition of compensation of the external action by the system, according to which the change in the distribution function for filled pores due to the external action is compensated by the response of the system in the form of flow of the liquid to an infinitely large cluster of empty pores. Analysis of Eq. (31) taking into account the change in the distribution function for accessible pores will be carried out below when we will consider oscillating modes of infiltration.

Equation (30) makes it possible to determine fraction θ_0 of the pores for which infiltration of a porous media can be initiated. It follows from Eq. (30) that the following estimate is valid in the case considered here:

$$\frac{F}{\varepsilon}\tau_c^{-1} = \frac{\lambda_\infty(\theta)}{\tau_0(p)}F \qquad (32)$$

Relations (26), (30), and (32) lead to the conclusion that the value of θ_0 for which infiltration of the porous media may begin is defined by the relation

$$(\varepsilon(\theta_c))^{\frac{2}{3}}\frac{\tau_0(p)}{\tau_p} = z(\theta_0 - \theta_c)^\varsigma \qquad (33)$$

which shows that fraction θ_0 of pores is defined by rate τ_p^{-1} of the pressure growth. Since $\varepsilon^{\frac{2}{3}}\frac{\tau_0}{\tau_p} \ll 1$, the value of θ_0 is close to θ_c. Pressure p_0 corresponding to the beginning of infiltration is defined by the relation analogous to (23):

$$\int_0^\infty w(R, p_0)dRf_r(R)R^3 = \theta_0 \qquad (34)$$

Thus, Eq. (33) makes it possible to find the fraction of pores for which the system compensates the external action, and infiltration of the liquid in the porous media begins, leading to a macroscopic change in the volume of the liquid in the porous media

The change in the volume of the system consisting of a porous media and a liquid occurs over time intervals $t \sim \tau_v > \tau_z \gg \tau_d$ due to infiltration of the liquid into an infinitely large cluster of accessible pores through finite-size clusters of filled pores. We will derive an equation describing the time dependence of the fraction of volume of pores filled with the liquid in the vicinity of θ_0. For this, we write distribution function $F(n,t)$ in the form

$$F(n,t) = x(t)F_1(n,t). \qquad (35)$$

Here, quantity $x(t)$ varies over time intervals $t \sim \tau_v \gg \tau_z$, while the variation of $F_1(n,t)$ occurs over time intervals $t \sim \tau_z \ll \tau_v$. Since the new stationary state appears for $\theta_c > \theta_{c0}$, we will calculate the filled volume assuming that the space of accessible pores is homogeneous and all pores of the porous media are accessible to infiltration. Using relation (9), we normalize the distribution functions for accessible and filled pores to the total volume of accessible pores, assuming that it is equal to unity. This is due to the fact that the value of p_0 increases with the compression energy (see below) so that all pores in the porous media become accessible and $\theta \to 1$. It follows from Eq. (9) that with such a normalization, all pores accessible to infiltration (including those belonging to the infinitely large cluster) are taken into account. In this case, quantity $x(t)$ is the fraction of filled pores at instant t. Assuming that function $F_1(n,t)$ is normalized to unity ($\sum_{n=1}^{\infty} nF_1(n,t) = 1$), we can represent the distribution function for accessible pores in the form

$$f(n) = (1-x(t))f_0(n, \varepsilon = \theta - \theta_c), \qquad (36)$$

$$\sum_{n=1}^{\infty} nf_0(n, \varepsilon = \theta - \theta_c) = 1$$

Substituting expressions (35) and (36) into Eq. (4) and considering that, by virtue of (31), the value of $\varepsilon = \theta_0 - \theta_c = const$ over time intervals $t \sim \tau_v > \tau_z \gg \tau_d$ (and, hence, $\dfrac{\partial F_1}{\partial \varepsilon} = 0$), we obtain

$$F_1 \frac{dx}{dt} + x\frac{\partial F_1}{\partial t} = \frac{x(1-x)}{\tau_0(p)} \lambda_\infty(\theta_0) F_1(n) + \frac{x(1-x)}{\tau_0(p)} \sum_m \tilde{A}_{nm} F(m) \qquad (37)$$

Consequently, for $t \sim \tau_v \gg \tau_z$, using expression (30), we obtain from relation (37).

$$\frac{dx}{dt} = \frac{x(1-x)}{\tau_v}, \tau_v = \frac{\tau_0(p)}{\lambda_\infty(\theta_0)} \tag{38}$$

Using relations (26) and (33), we obtain the characteristic volume infiltration time

$$\tau_v = \frac{\tau_0(p)}{\lambda_\infty(\theta_0)} = (\varepsilon(\theta_0))^{-\frac{2}{3}} \tau_p \tag{39}$$

It follows hence that if condition (30) describing the compensation of an external action by the system is satisfied, characteristic volume infiltration time τ_v is controlled by characteristic time τ_p of pressure growth and by the difference $\theta_0 - \theta_c$; consequently, it is independent of the viscosity of the liquid.

Thus, in the case of fast variation of pressure $(\tau_z > \tau_p \gg \tau_d)$, infiltration of the porous media occurs via rapid infiltration of finite-size clusters, occurring simultaneously in the entire volume of the grain (over time intervals $t \sim \tau_z$) and slow infiltration (over time intervals $t \sim \tau_v \gg \tau_z$) of the liquid into the infinitely large cluster of accessible pores through finite-size clusters of filled pores. As a result, the new state of the system consisting of the nonwetting liquid and the porous media is formed over time periods $t < \tau_v$ as a result of the nonlinear response of the system to the external action. Equation (38) describing the infiltration of a porous media was proposed phenomenologically in [27].

3.5. Oscillating Modes of Infiltration

System of equations (4), (5) for $\theta \geq \theta_c$ may have solutions oscillating with time. To prove this formally, we will seek the solution to system (4), (5) for $\theta \approx \theta_0$ in the form

$$F(n,t) = F_0(n) + \delta F(n,t), |\delta F(n,t)| \ll F_0;$$
$$f(n,t) = f_0(n) + \delta f(n,t), |\delta f(n,t)| \ll f_0; \quad (40)$$

Here, $f_0(n)$ and $F_0(n)$ are defined by formulas (10) and (12),

$$F_0(n) = \frac{Y\Omega_n(t)}{Z(t)}, f_0(n) = \frac{(\theta_c - Y)\Omega_n(t)}{Z} \quad (41)$$

where Y is the fraction of filled pores. Substituting expressions (40) into Eqs. (4) and (5) and using the tau-approximation, in the first order in deviations $\delta F(n,t)$ and $\delta f(n,t)$, we obtain

$$\delta \dot{f}(n,t) = -\frac{f_0(n)}{\tau_d}\delta f(n,t) + \frac{F_0(n)}{\tau_z}\delta f(n,t) + \frac{f_0(n)}{\tau_z}\delta F(n,t) + \frac{S(\varepsilon)}{\tau_{pc}}\delta F(n,t)$$
$$\delta \dot{F}(n,t) = -\frac{F_0(n)}{\tau_z}\delta f(n,t) - \frac{f_0(n)}{\tau_z}\delta F(n,t) - \frac{S(\varepsilon)}{\tau_\infty}\delta F(n,t), \quad (42)$$

where $\varepsilon = \theta_0 - \theta_{c0}$. The equation for $\delta \dot{F}(n,t)$ corresponds to Eq. (30) written in the x approximation. System of linear equations (42) has oscillating solutions if the eigenvalues of the matrix of this system contain imaginary parts. This is the case when the condition

$$(1 - \frac{A}{B})^2 - 2\frac{G}{B} - 2\frac{AG}{B^2} + \frac{G^2}{B^2} < 0 \quad (43)$$

is satisfied, where

$$A = \frac{F_0(n)}{\tau_z}, B = \frac{f_0(n)}{\tau_d}, G = \frac{f_0(n)}{\tau_z} + \frac{S(\varepsilon)}{\tau_\infty}$$

It follows from relation (43) that the condition for the emergence of oscillations can be satisfied only when $A \neq 0, B \neq 0$, and $G \neq 0$. Assuming that $\tau_{pc} \sim \tau_z$ and using relation (43), we obtain the following condition for the existence of oscillating modes of infiltration of a porous media:

$$W(x',Y,\theta,n) = (1-x'\frac{Y}{\theta-Y})^2 - 2x'(1+\frac{S(\varepsilon)}{f_0(n)}) - 2x'^2\frac{Y}{\theta-Y}(1+\frac{S(\varepsilon)}{f_0(n)}) + x'^2(1+\frac{S(\varepsilon)}{f_0(n)})^2 < 0 \quad (44)$$

where $x' = \tau_d/\tau_z$. For preset values of the total fraction θ of filled and accessible pores, fraction Y of filled pores, and ratio of times $x' = \tau_d/\tau_z \ll 1$, inequality (44) defines the number of pores constituting the cluster whose infiltration-defiltration results in the emergence of an oscillating mode. The solution of linear system of equations (42) is cumbersome and is not given here. Characteristic period T_0 of oscillations coincides in order of magnitude with the infiltration-defiltration characteristic time for a cluster of filled pores. Analysis of condition (44) shows that in the case of slow infiltration $(\tau_v > \tau_p \gg \tau_z > \tau_d)$, when $\theta \leq \theta_{c0}$ and $S(\varepsilon) = 0$, oscillations are absent. In the case of rapid infiltration $(\tau_v > \tau_z > t \geq \tau_p > \tau_d)$, in the vicinity of the transition to the new state, we have $\theta \sim \theta_0$, $S(\varepsilon) > 0$, and the fraction of filled pores is small $(Y \ll 1)$. In this case, condition (44) can be satisfied for a certain value of number n.

Figure 9 shows a typical graph of function $W(x',Y,n)$ and eigenvalues of the matrix of Eqs. (42) as functions of number of pores n in the cluster. It can be seen that the real parts of the eigenvalues of the matrix of system (42) are negative for any number of pores in the cluster. Condition (44) is satisfied for a cluster consisting of $n < n_k = 2$ pores; for $n < 2$, the eigenvalues of the matrix of system of equations (42) acquire an imaginary part that vanishes for $n \geq 2$ (see Figure 9).

The existence of the imaginary part in the eigenvalues of the matrix of system (42) corresponds to the emergence of an oscillating infiltration mode. Consequently, for $x' = 0.01$, $Y = 0.01$, and $\theta_0 = 0.28$, an oscillating mode appears as a result of infiltration-defiltration of a cluster containing $n \sim 2$ pores. It follows from expression (6) that this process occurs over time intervals of approximately $2\tau_0$; consequently, when the conditions of fast infiltration are satisfied, oscillations with a period $T_0 \sim 2\tau_0$ may appear and will accompany infiltration of the porous media.

Substituting Eq. (40) into (9), taking into consideration that condition (44) is satisfied for $n \sim n_k \approx 2$, and retaining the oscillating part, we obtain

$$\theta(t) = \theta_0 + \delta\theta(t), |\delta\theta(t)| << \theta_0$$
$$\delta\theta(t) = n_k(\delta f(n_k,t) + \delta F(n_k,t))$$
(45)

For pressure oscillations, we obtain from Eq. (3)

$$\delta p(t) = \delta\theta(t)(\int_0^\infty \frac{\partial w(R, p_c)}{\partial p} dR f_r(R) R^3)^{-1}$$
(46)

Thus, the emergence of oscillations $\delta f(n_k,t)$ and $\delta F(n_k,t)$ of the distribution functions leads to the emergence of pressure oscillations with a period $T_0 \sim 2\tau_0$.

Let us find the time dependence of the volume in an oscillating regime. For this purpose, we substitute Eqs. (45) into Eq. (38). The solution of the resultant equation has the form

$$x(t) = \frac{\theta_0}{1 + \frac{\theta_0 - x(0)}{x(0)} \exp(-\frac{1}{\tau_v} \int_0^t (\theta_0 + (1 + \varsigma(\theta_0 - \theta_c)^{s-1})\delta\theta(t)) dt)}$$
(47)

Here, $x(0)$ is the fraction of pores filled by the instant of passage of the system to a new state and quantity τ_v is defined by relation (31). Since the characteristic period of oscillations is $T_0 \sim 2\tau_0 << \tau_v$, the value of $\int_0^t \delta\theta(t)dt \approx 0$ over time intervals $t \geq \tau_v$ and, hence, oscillations of the relative volume must be smaller than pressure oscillations (46).

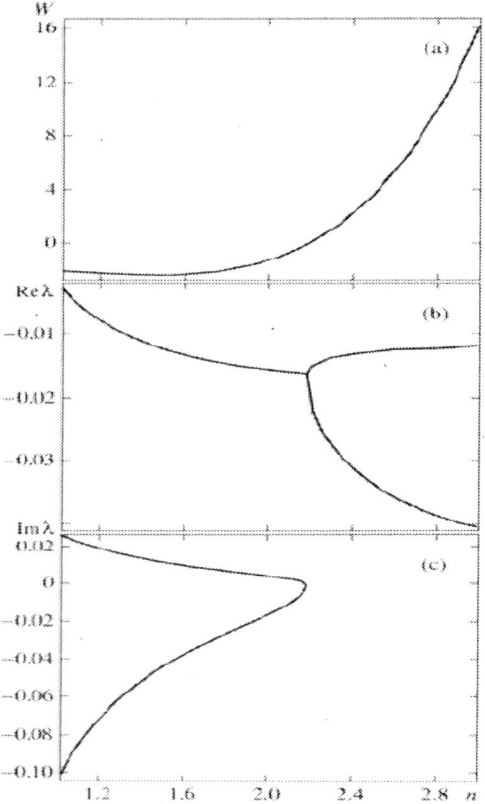

Figure 9. Dependences (a) of function $W(x', Y, n)$, as well as (b) of real and (c) imaginary parts of eigenvalues of matrix (44) on number n of pores in a cluster, plotted for $x' = 0.01$, $Y = 0.01$, and $\theta_0 = \theta_c = 0.28$.

In our experiments, a liquid Wood's alloy ($T_m = 72°C$) and a Silokhrom SKh-1.5 porous media were placed in a high-pressure chamber. The mass and size of the Silokhrom grains were $m = 1$ g and 300 µm, respectively. The pore diameter in Silokhrom SKh-1.5 ranged from 130 to 260 nm. The pressure in the chamber was produced by mechanical action on a rod that could enter the chamber through gaskets. A decrease in volume of the Wood's alloy-Silokhrom system upon moving the rod inside the chamber produced excessive pressure. The change in volume was measured using a displacement pickup. The pressure was measured by a strain gauge that was mounted on a support under the high-pressure chamber. The gauge could detect strength from 0 to 1000 kg in the frequency range up to 10 kHz with an accuracy of=10%. The filling critical pressure was determined from the $V(p)$ dependence of the filled pore volume on pressure p for a quasi-static pressure buildup with a characteristic time of=10 s. For the system under study, this value was found to be p_c = 120 atm. In the experiments with dynamic filling, the time-dependent pressure in the chamber was measured for the pulsed mechanical action on the chamber rod. The measured compressibility of Silokhrom SKh-1.5 was χ = 1.6 x 10^{-3} atm^{-1}. Since the compressibility of the chamber with volume V_{ch} =120 cm^3 was χ = 1.4 x 10^{-5} atm^{-1}., a change in the chamber volume filled with the Wood's alloy (the compressibility of the Wood's alloy is ~10^{-6} atm^1) was vastly larger than the change in the Silokhrom volume in the dynamic experiments on the time scale of <10 ms with working pressure p ~ 3 x 0^2 atm. Because of this, the characteristics of a pressure pulse in the chamber were determined in special experiments, in which the chamber was filled only with the Wood's alloy. When studying the filling dynamics of the porous media, the maximal pressure in the chamber was p_0 = 240-600 atm, i.e., much higher than the critical pressure of the system of interest. The characteristic time t_1 of reaching the maximal pressure was varied in these experiments within 4-11 ms, and the characteristic time of pressure release was 5-10 ms.

The p(t) curves for the pressure in the chamber filled with the Wood's alloy and porous media is shown in Figure 10b.

The corresponding $p_0(t)$ curves for the chamber filled only with the Wood's alloy is shown in Figure 10a. For the short p_0 pulse (p_0 max = 450 atm, upper panel in Figure 10a), periodic oscillations with the characteristic period T~1 ms and amplitude δp ~ 20 ± 2 atm appear in the $p(t)$ curve (upper panel in Figure 10b). It is seen from the middle panel in Figure 10b that, at a fixed duration, an increase in the amplitude of pulse p_0 (middle panel in Figure 10a) gives rise to the additional harmonics in the $p(t)$ dependence. As the p_0 pressure amplitude decreases and the pulse duration increases (lower panel in Figure 10a), the oscillations in the $p(t)$ curve disappear (lower panel in Figure 10b). The instants of time t_2 (Figure 10b) corresponding to the completion of filling the porous media with the Wood's alloy were determined from the momentum conservation law. One can see that the oscillations are observed at $t < t_2$. At $t > t_2$, the liquid leaks away from the porous media. At these times, the $p(t)$ curves also display oscillations (upper and middle panels in Figure 10b). It follows from the data in Figure 10 and from the additional experiments that, at a fixed duration of the p_0 pulse, there is a critical pressure p_{0c} = 300 atm below which the filling oscillations are absent. Note that the increase in the p_0 pulse duration from 10 to 20 ms also results in the disappearance of oscillations.

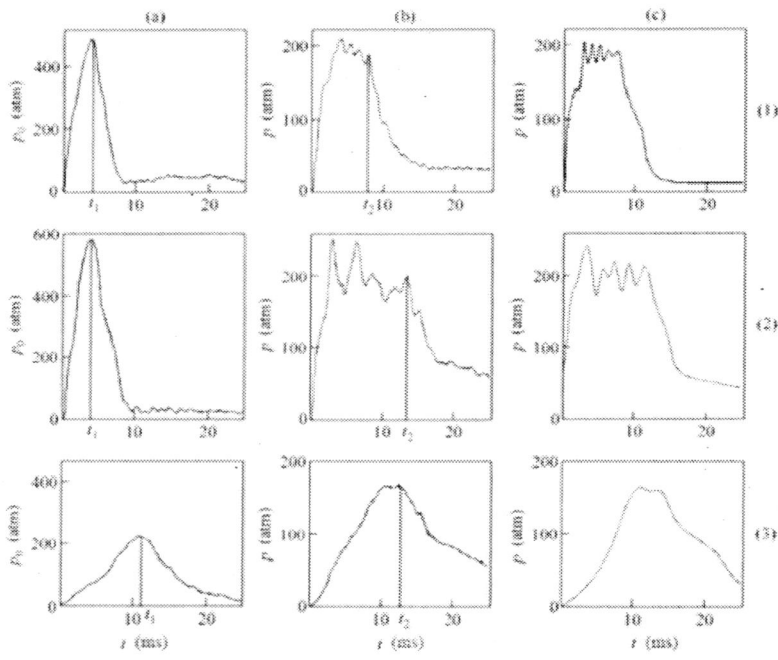

Figure 10. Plots of pressure in the chamber vs. time: (a) a column filled with a liquid Wood's alloy; (b) a column filled with a liquid Wood's alloy and a porous media (Silokhrom SKh-1.5, 1 g); and (c) numerical simulation. The panels correspond to different values of $p_{0\ max}$ and t (see text).

3.6. Physical Pattern of Infiltration of a Nonwetting Liquid into a Porous Media

Thus, in accordance with the model considered here, we obtain the following pattern of infiltration of a nonwetting liquid through pores of a disordered porous media (see Figure 7). Infiltration is described as a spatially nonuniform process with the help of distribution functions $f(n,t)$ and $F(n,t)$ for clusters of accessible empty and filled pores, respectively. These functions satisfy kinetic equations taking into account the pair "interactions" of accessible-pore clusters with clusters of accessible filled pores. In the case of a slow infiltration, for which characteristic time τ_p of pressure growth is longer than the characteristic infiltration time in the vicinity of percolation threshold θ_{c0} from the standard percolation theory, the critical retardation effect should be observed (see Figure 7). In this case, the characteristic infiltration time $\tau_v \to \infty$ for $\theta \to \theta_{c0}$ and, hence for pressure $p \to p_{c0}$. Under these conditions, all pores accessible at this pressure are being filled and the cluster of empty accessible pores is not formed ($P(\varepsilon) = 0$, see Figure 7).

Under a rapid pressure increase, in which time τ_p is shorter than characteristic time x_z of infiltration through clusters of accessible pores, these pores have no time to be filled before the attainment of percolation threshold θ_{c0} over time scale $t \sim \tau_z$, so that values of $t \sim \tau_z$ are attained and an infinitely large cluster of accessible but empty pores is formed. In this case, $P(\varepsilon) \neq 0$ (see Figure 7). Steady-state distribution functions $f_0(n)$ and $F_0(n)$ are formed over time intervals τ_d and τ_z, respectively, such that $\tau_d < \tau_z$. It follows from the solvability condition for the system of kinetic equations for function $F_0(n)$ that this function differs from zero only when the pressurized system attains a new infiltration threshold in the fraction of empty accessible pores $\theta_c = 0.28$), which is higher than the familiar percolation threshold $\theta_{c0} = 0.18$. The threshold value of $\theta_c = 0.28$ is a new characteristic of the dynamics of infiltration of the porous media. In the range of θ from 0.18 to 0.28, infiltration of the porous media should not be observed over a characteristic time of pressure growth $\tau_p < \tau_z$ (see Figure 7).

Analysis shows that for $t \gg \tau_z$, distribution function $f(n,t)$ for clusters of accessible pores is quasistationary, $f(n,t) \approx f_0(n, \varepsilon(t))$, while the kinetic equation for distribution function $F(n,t)$ for clusters of filled pores for $n \to \infty$ and $\theta > \theta_c$ has a small positive (in the vicinity of the new threshold) eigenvalue $\lambda_\infty \sim (\theta - \theta_c)^\varsigma$, $\varsigma \approx 0.8$. This eigenvalue controls the characteristic time of the increase in the macroscopic volume of the pore space filled with the liquid ($\tau_v \sim \lambda_\infty^{-1}$). Remaining eigenvalues λ_n are negative and correspond to characteristic relaxation frequencies for finite-size clusters of filled pores. Consequently, the infiltration of a grain of a porous media is a rapid (with characteristic time $\tau_z \ll \tau_v$) process of the formation of finite-size clusters of filled pores around an infinitely large cluster of accessible pores. The liquid flows through these clusters to the infinitely large cluster of accessible pores, filling it over a macroscopic time $t \geq \tau_v \gg \tau_z$.

The solution to the kinetic equation for distribution function $F(n,t)$ shows that the increase in distribution function $F(n,t)$ (and, hence in quantity θ) during infiltration time τ_v is compensated by the variation of $F(n,t)$ due to infiltration of clusters of accessible pores as a result of their interaction (percolation of the liquid) with clusters of filled pores, as well as due to percolation of the liquid to the infinitely large cluster of accessible pores from finite-size clusters of filled pores. Such a compensation can take place since the system consisting of the nonwetting liquid and the porous media is "thrown" beyond new infiltration threshold θ_c, and characteristic infiltration time τ_v for $\theta = \theta_0 > \theta_c$ is independent of the viscosity of the liquid (which ensures this dynamic compensation) and is determined by time τ_p of the pressure growth so that $\tau_v = \tau_p (\theta_0 - \theta_c)^{-\rho}, \rho \approx 0.4$. Thus, fraction θ_0 of accessible pores and, hence, the pressure remain unchanged during infiltration of a porous media.

For a fraction of accessible pores close to (or higher than) new threshold θ_0, the system of kinetic equations for the distributions functions $f(n,t)$ and $F(n,t)$ of clusters of accessible empty and accessible filled pores has oscillating solutions. The characteristic scale of the period of oscillations is on the order of τ_z. The oscillations must be observed during infiltration time τ_v and correspond to the periodic infiltration-defiltration of the liquid in finite-size clusters.

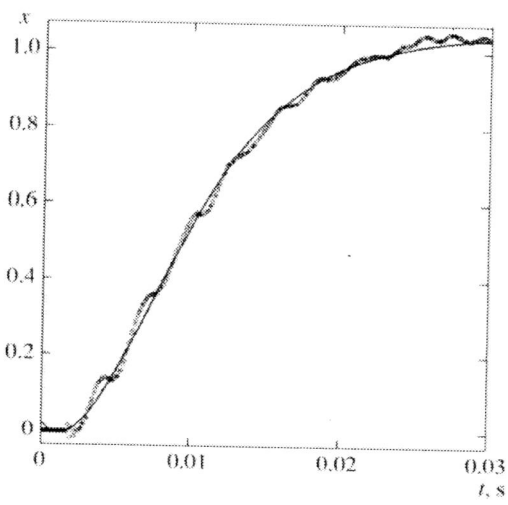

Figure 11. Time dependence of fraction x of filled volume for the L23 + $CaCl_2$ system (circles) and the corresponding curve calculated by Eq. (38) (solid curve).

Chapter 4

4. DISCUSSION OF RESULTS AND COMPARISON WITH EXPERIMENT

Experiments show that the characteristic times of variation of volume and pressure for the system of a nonwetting liquid and a nanoporous media considered here are $\tau_v \sim 25$ ms and $\tau_p \sim 5$ ms, respectively (see Figure 5). Let us estimate characteristic times τ_z and τ_0. For this purpose, we write

$$\tau_v \sim \frac{V(L)}{SJ}, \qquad (48)$$

where $V(L)$ is the volume of a grain of size L, S is the surface area of the grain through which the liquid infiltrates, and J is the flow rate of the liquid. Using the linear relation between the velocity of the liquid flow and pressure gradient,

$$J = \frac{k_n}{\eta} \frac{\Delta p}{L}, \qquad (49)$$

where k_n is an unknown coefficient having the meaning of the penetrability factor of the medium [28], and assuming that $V(L) = \frac{4\pi}{3} L^3$ and $S = 4\pi L^2$, we obtain

$$\tau_v \sim \frac{L^2 \tilde{\eta}}{3 k_n \Delta p} \qquad (50)$$

Assuming that time τ_v is given, we can obtain unknown coefficient k_n from relation (49):

$$k_n \sim \frac{L^2 \tilde{\eta}}{3 \tau_v \Delta p} \qquad (51)$$

Then definition (6) for $\tau_0 = \dfrac{4 \tilde{\eta} \overline{R} L}{3 k_n \Delta p}$ gives

$$\tau_0 \sim \frac{4 \overline{R}}{L} \tau_v \qquad (52)$$

For $\overline{R} \sim 10$ nm and $L \sim 1$ μm, the value of τ_0 is approximately 1 ms. Using relation (6) for hydrodynamic time $\tau_z \sim \tau_0(p)\langle n^{-q} m^{1-q} \rangle$ for $\theta \sim \theta_0 \sim 0.3$, we obtain $\tau_z \sim 10$ ms. Using relation $\tau_d = \varepsilon(\theta_0)^{1+\gamma} \tau_p$, for $\gamma = 0.6$, we obtain $\tau_d \sim 0.1$ ms. Thus, the inequalities typical of rapid infiltration ($\tau_v > \tau_z > \tau_p > \tau_0 > \tau_d$) are observed in experiments with the systems under investigation, which allows us to use the developed model of infiltration for interpreting experimental data. For a slow variation in pressure ($\dot{p} \sim 1$ atm/s), $\tau_p \sim 100$ s and, hence $\tau_p \gg \tau_z > \tau_d$. This justifies the use of relations (10) and (12) for describing the results of experiments on slow infiltration of a nonwetting liquid through a nanoporous media [8, 20].

We can estimate the value of quantity θ_0 emerging due to compensation of the external action by the nanoporous media being infiltrated and determining the infiltration threshold using relation (33), which gives

$$\theta_0 = \theta_c + (\frac{\tau_0}{z\tau_p}\varepsilon(\theta_c)^{\frac{2}{3}})^{\frac{1}{\zeta}} \tag{53}$$

for $\tau_0 \sim 1$ ms and $\tau_v \sim 25$ ms, we obtain $\theta_0 = \theta_c + 0.003 \approx 0.283$. Thus, infiltration of a porous media in these experiments may begin in the vicinity of the transition of the system to a new infiltration state for $|\theta_0 - \theta_c|/\theta_c \sim 10^{-2}$. It should be borne in mind that in view of the finite size of grains, percolation threshold θ_{c0} (and, hence, quantities such a θ_0 and θ_c determined from it) also have values differing from the values of the corresponding quantities in the case of an infinitely large medium.

This difference can be estimated assuming that infiltration through a grain of size L begins when correlation length $\varsigma \sim \dfrac{\overline{R}}{\varepsilon^\nu(\theta)}$, $\nu \approx 0.8$ [32] becomes equal to the grain size. Consequently,

$$\theta_{c0}(L) = \theta_{c0} - (\frac{\overline{R}}{L})^{\frac{1}{\nu}} \tag{54}$$

This gives $\theta_{c0}(L) - \theta_{c0} \sim 10^{-3}$, indicating the possibility of using the infinitely large medium approximation for describing the result of experiments.

It follows from relations (22) and (23) that positive eigenvalue $\lambda_\infty > 0$, which determines the characteristic infiltration time of the porous media, is formed in the limit $n \to \infty$. The numerical solution of Eqs. (4) and (5) shows that the formation of a positive eigenvalue of system of equations (4), (5) begins for a number of pores $n > 100$ in the cluster. At the same time, a grain of size $L \sim 1-10$ μm in the system under investigation contains a number of pores $n \sim \dfrac{L^3}{\overline{R}^3} \sim 10^6 - 10^9$, which allows us to use the relations obtained in Section 3 in the infinitely large medium approximation for describing experimental dependences.

Infiltration of a porous media occurs for the fraction $\theta_0 \approx 0.28$ of accessible pores. For the systems under investigation, the infiltration pressure is $p_0 \approx 1.2 p_{c0}$. It follows from expression (1) that the probability of a pore being accessible to infiltration is 0.93; consequently, 93% of all pores in a grain of the porous media become accessible to infiltration at room temperature. It can be seen from Figure 6 that for $\theta_0 \approx 0.28$, about 70% of pores in the porous media belong to the infinitely large cluster of accessible pores. The remaining 23% of pores (that do not belong to the infinitely large cluster) form finite-size clusters. These clusters surround the infinitely large cluster and, being infiltrated over time interval $\tau_z \sim 10$ ms, form finite-size clusters of filled pores through which the liquid flows to the infinitely large cluster of accessible pores, infiltrating it over the characteristic time $\tau_v \sim 25$ ms. Thus, infiltration of a grain of a rapidly pressurized porous media can be treated as a uniform process of infiltration of the liquid though finite-size clusters of accessible pores (about 20% of pores in the entire porous media) occurring simultaneously in the entire space of pores in the grain with characteristic time τ_z; it is followed by percolation of the liquid from these clusters to the growing infinitely large clusters of accessible pores, containing about 70% of all pores in the porous media with a long characteristic time $\tau_v \sim 25$ ms.

Our experiments show that the time dependences of infiltration pressure and filled volume do not change upon a fivefold change in the viscosity of the liquid. Infiltration pressure p_c at which the porous media passes to a new state can be determined from Eq. (34), which does not contain viscosity; consequently p_0 is independent of viscosity. Numerical solution of Eq. (34) for $\theta_0 = 0.28$ gives $p_0 = 200$ atm, which corresponds to experimental data (see Figure 5). It follows from Eq. (39) and relation (40) that the volume infiltration time is independent of the viscosity of the liquid due to dynamic compensation of the external action by the nonwetting liquid-nanoporous media system. Figure 11 shows the time dependence of fraction x of the filled volume for the L23 + $CaCl_2$ system, calculated using Eq. (38). It can be seen that the experimental and theoretical dependences almost coincide. It should be noted that Eq. (38) is valid over time intervals $t \sim \tau_v > \tau_z \sim 10$

Discussion of Results and Comparison with Experiment 57

ms. Consequently, the coincidence of the theoretical and experimental time dependences of the filled volume at shorter times is accidental.

In the model developed here, the filled volume and the infiltration time of a porous media are functions of the compression energy. These dependences can be derived from Eq. (38). Indeed, multiplying Eq. (38) by p_0 and taking into account that $E = \int_0^{x_{max}} p dx$, where x_{max} is the maximum fraction of filled volume, we obtain

$$E = \int_0^{x_{max}} p_0 dx = \int_0^{\tau_{in}} dt \frac{p_0}{\tau_v} x(1-x) \approx \frac{p_0}{2\tau_v} \tau_{in} \qquad (55)$$

Pressure depends on the compression energy only slightly and has a tendency to increase within the experimental error (see Figure 3a). Since p_0 and τ_v are independent of the compression energy in the zeroth approximation, relation (55) shows that infiltration time $\tau_{in} \sim E$. Integrating Eq. (38) with respect to time, we obtain

$$x_{max} = \int_0^{\tau_{in}} \frac{dt}{\tau_v} x(1-x) \approx \frac{\tau_{in}}{2\tau_v} \propto E \qquad (56)$$

It follows hence that in the model developed here, the maximum filled volume is a linear function of the compression energy; consequently, the flow rate of the liquid during infiltration of the porous media is independent of energy. Dependences (55) and (56) for the infiltration time and for the maximum filled volume on the compression energy describe the experimental data to within the measurement error (see Figures 3b and 3c).

The value of pressure p_0 corresponding to the beginning of infiltration also depends on energy. Indeed, Eqs. (29) and (33) lead to

$$(\theta_0 - \theta_c)^\varsigma \sim (\frac{\tau_0(p_c)}{\varepsilon \tau_p} p_c (\frac{\partial \varepsilon}{\partial p})_{p=p_c}), \; \varsigma \approx 0.8 \qquad (57)$$

Integrating this equation by p, we obtain

$$\int dp p_c \frac{\tau_0(p_c)}{\tau_p}(\frac{\partial \varepsilon}{\partial p})_{p=p_c} \approx \frac{\tau_0(p_c)}{\tau_p}\int p_c d\theta \approx \frac{\tau_0(p_c)}{\tau_p}E \qquad (58)$$

$$\int dp(\theta_0 - \theta_c)^\varsigma = (\theta_0 - \theta_c)^\varsigma p_c \qquad (59)$$

Here, E is the compression energy per unit volume of the nanoporous media. Using expressions (58) and (59), we obtain from Eq. (57)

$$\theta_0(E) = \theta_c + (\frac{\tau_0(p_c)}{\tau_p}\frac{E}{p_c})^{\frac{1}{\varsigma+1}} \qquad (60)$$

Suppose $p_0(E) = p_c + \delta p(E)$, we obtain from Eq. (34)

$$\delta p(E) = (\int_0^\infty \frac{\partial w(R, p_c)}{\partial p}dRf_r(R)R^3)^{-1}(\frac{\tau_0(p_c)}{\tau_p}\frac{E}{p_c})^{\frac{1}{\varsigma+1}} \qquad (61)$$

It follows hence that $\delta p(E) \propto E^{1/(\varsigma+1)}$ and therefore $p_0(E) - p_c \propto E^{1/(\varsigma+1)}$. Consequently, when the external action with various compression energies is compensated, the attained excess over threshold value of θ_0 and, hence, the excess of infiltration pressure p_0 over the threshold pressure depend on the compression energy. This ensures an infiltration rate (with characteristic time τ_v) satisfying relation (33). The dependence of pressure p_0 on the compression energy is shown in Figure 3a. It can be seen that experimental data are described by relation (61) correct to the measurement error. In accordance with Eq. (32), reflecting the condition of compensation of the external action by the system, the time dependences of the infiltration pressure and filled volume in our model do not change with the viscosity of the liquid for this system (see Figure 5).

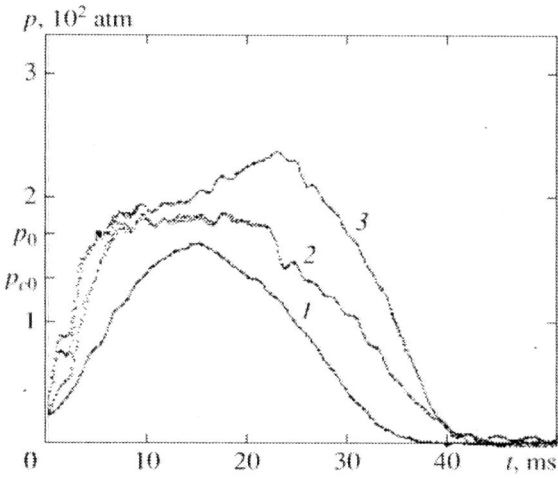

Figure 12. Time dependences of infiltration pressure p for the L23 + H$_2$O system for various compression energies E = 10 (*1*), 35 (*2*), and 60 J (*3*). The mass of the porous media is 4 g and the threshold pressures are $p_{c0} = 150$ atm and $p_c = 180$ atm.

Thus, for systems L23 + H$_2$O and L23 + CaCl$_2$, infiltration of the porous media under rapid compression is a nonuniform process in which clusters of filled pores ensuring percolation of the liquid to an infinitely large cluster of accessible but empty pores are formed after a new threshold θ_c of infiltration through accessible pores is attained that exceeds the known percolation threshold θ_{c0}. In accordance with the proposed model, the absorption of the compression energy occurs at a constant pressure p_0, which exceeds pressure p_c corresponding to the new threshold θ_c. This pressure p_0 is determined by the condition of compensation of the external action by the increase in the number and size of clusters of filled pores at a given rate of pressure growth, which ensures percolation of the liquid to the infinitely large cluster of accessible pores at a constant pressure. Such an infiltration regime takes place if characteristic time τ_p of pressure growth is shorter than the characteristic time of infiltration of the cluster of accessible pores. This is ensured (under an increase in pressure) by the attainment of value $\theta > \theta_c$ for the fraction of accessible pores without infiltration of accessible pores and by the formation of an infinitely large cluster of accessible pores.

On the other hand, when the compensation condition is satisfied, the value of τ_p may be smaller than characteristic infiltration time τ_v or close to it, in accordance with Eq. (39), which contains factor $\varepsilon^{-2/3}(\theta_0)$. In this pattern, if pressure p is (as a result of rapid pressurization) such that the inequality $p_{c0} < p < p_c$ is satisfied, infiltration does not take place in the porous media. The existence of such an infiltration mode (the presence of a pressure "gap" in infiltration) is confirmed by the experimental results depicted in Figure 12. It can be seen that the maximum pressure $p = 160$ atm was attained during the infiltration of the porous media (curve 1). This value is higher than $p_{c0} = 150$ atm, but lower than threshold $p_0 = 180$ atm. To within the experimental error, the change in the volume of the system coincides with the change in the volume due to deformation. The infiltration regime at a constant pressure is not observed if the compression energy exceeds the maximum value ($E = 60$ J; curve 3 in Figure 12) determined by the specific energy of infiltration of the porous media. In these conditions, the decrease in the flow rate of the liquid during infiltration of the porous media may attain a value of $J < J_{min} = J(E) = const$ and the required energy absorption rate is not ensured for the characteristic time of pressure growth. For this reason, the response of the system consisting of a nonwetting liquid and a nanoporous media is an increase in pressure to the maximum value, followed by defiltration (curve 3 in Figure 12). It should be noted that the value of J_{min} cannot be calculated in the mean field approximation under the assumption of invariability in the medium during infiltration since initial kinetic equations (3) and (4) disregard the interaction of clusters of filled pores.

ACKNOWLEDGMENTS

The authors are grateful to L.A. Maksimov, who read the manuscript and made some valuable remarks, and to I.V. Tronin for fruitful discussions and assistance.

REFERENCES

[1] Sahimi, M. (1993). *Rev. Mod. Phys.*, *65*, 1393.
[2] Thompson, A. H., Katz, A. J. & Raschke, R. A. (1987). *Phys. Rev. Lett.*, *58*, 29.
[3] Feder, J. (1988). *Fractals* (Plenum, New York, Mir, Moscow, 1991).
[4] Bogomolov, V. N. (1978). Usp. Fiz. Nauk, *124*, 171, [Sov. Phys-Usp. 21, 77].
[5] Yu, A. Fadeev, & Eroshenko, V. A. (1995). *Ross. Khim. Zh.*, *39(6)*, 93,
[6] Eroshenko, V. A. & Yu. A. (1996). *Fadeev, Zh. Fiz. Khim.*, *70(8)*, 1482 [Russ. *J. Phys. Chem. 70(8)*, 1380].
[7] Matthews, G. P., Ridgway, C. J. & Spearing, M. C. (1995). *J. Colloid Interface Sci.*, 171, 8.
[8] Kloubek, J. (1994). *J. Colloid Interface Sci.*, *163*, 10.
[9] Borman, V. D., Grekhov, A. M. & Troyan, V. I. (2000). *Zh. Éksp. Teor. Fiz.*, *193*, [JETP 91 (1), 170 (2000)].
[10] Kong, X., Surani, F. B. & Qiao, Y. (2005). *J. Mater. Res.*, *20*, 1042.
[11] Qiao, Y. & Kong, X. (2005). *Phys. Scr.*, *71*, 27.
[12] Surani, F. B., Kong, X. & Qiao, Y. (2005). *Appl. Phys. Lett.*, *87*, 251906.
[13] Qiao, Y. & Kong, X. (2005). *Phys. Scr.*, *71*, 27.
[14] Kong, X. & Qiao, Y. (2005). *Appl. Phys. Lett.*, *86*, 151919.
[15] Suciu, C. V., Iwatsubo, T. & Deki, S. (2003). *J. Colloid Interface Sci.*, *259*, 62.
[16] Han, A., Kong, X. & Qiao, Y. (2006). *J. Appl. Phys.*, *100*, 014308.
[17] Surani, F. B. & Qiao, Y. (2006). *J. Appl. Phys.*, *100*, 034311.
[18] Surani, F. B., Han, A. & Qiao, Y. (2006). Appl. *Phys. Lett.*, *89*, 093108.

[19] Chen, X., Surani, F. B., Kong, X. (2006). et al., *Appl. Phys. Lett.*, *89*, 241 918.
[20] Kong, X. & Qiao, Y. (2006). *J. Appl. Phys.*, *99*, 064 313.
[21] Borman, V. D., Belogorlov, A. A., Grekhov, A. M., Lisichkin, G. V., Tronin, V. N. & Troyan, V. I. (2005). *Zh. Éksp. Teor. Fiz.*, *127(2)*, 431 [JETP 100 (2), 385].
[22] Nielsen, L. E. & Lande, R. F. (1993). *Mechanical Properties of Polymers and Composites* (Marcell Dekker, New York,).
[23] Bogomolov, V. N. (1995). *Phys. Rev. B: Condens.* Matter, *51*, 17040.
[24] Borman, V. D., Belogorlov, A. A., Grekhov, A. M., Lisichkin, G. V., Tronin, V. N. & Troyan, V. I. (2004). *Pis'ma Zh. Tekh. Fiz.*, *30(23)*, 1 [Tech. Phys. Lett. 30 (12), 973 (2004)].
[25] Shklovskii, B. I. & Efros, A. L. (1979). *Electronic Properties of Doped Semiconductors* (Nauka, Moscow, Springer, Berlin, 1984).
[26] Isichenko, M. B. (1992). *Rev. Mod. Phys.*, *64*, 961.
[27] Surani, F. B., Kong, X., Panchal, D. B. & Qiao, Y. (2005). *Appl. Phys. Lett.*, *87*, 163 111.
[28] Borman, V. D., Belogorlov, A. A., Grekhov, A. M., Tronin, V. N. & Troyan, V. I. (2001). *Pis'ma Zh. Éksp. Teor. Fiz.*, *74(5)*, 287, [JETP Lett. 74 (5), 258 (2001)].
[29] Basniev, K. S., Kochina, I. N. & Maksimov, V. M. (1993). *Underground Hydromechanics* (Nedra, Moscow) [in Russian].
[30] *The Chemistry of Grafted Surface Compounds*, Ed. by G. V. Lisichkin (Fizmatlit, Moscow, 2003) [in Russian].
[31] *Handbook of Physical Quantities*, Ed. by In: I. S. Grigoriev, & E. Z. Meilikhov, (Énergoatomizdat, Moscow, 1991; CRC Press, Boca Raton, FL, United States, 1997).
[32] *Handbook of Chemistry and Physics*, Ed. By In: D. R. Lide, (CRC Press, London, 1994).
[33] Yu. Yu. Tarasevich, Percolation: Theory, Applications, and Algorithms (URSS, Moscow, 2002).
[34] Abrikosov, A. A. (1979). Pis'ma Zh. Éksp. Teor. Fiz. 29 (1), 72 (1979) [JETP Lett. 29 (1), 65].
[35] Morse, P. M. & Feshbach, H. (1953). *Methods of Theoretical Physics* (McGraw-Hill, New York, Inostrannaya Literatura, Moscow), *Vol. 2*.

INDEX

A

alloys, 2
amplitude, 11, 19, 51
aqueous solutions, 9
attachment, 27, 28
averaging, 30, 40

B

behavior, 35

C

clusters, 1, 3, 4, 5, 6, 22, 24, 25, 26, 27, 28, 29, 30, 37, 40, 42, 44, 45, 52, 53, 54, 58, 61, 63
compensation, 6, 42, 45, 54, 57, 59, 61, 62
compliance, 4
compressibility, 11, 50
compression, vii, 1, 3, 4, 5, 6, 9, 11, 13, 15, 16, 17, 18, 44, 59, 60, 61, 62
connectivity, 22
conservation, 29, 30, 51
control, 18
correlation, 3, 57
critical value, 21

D

definition, 56
deformation, 62
displacement, 9, 50
distribution, 1, 2, 3, 5, 6, 16, 18, 25, 26, 27, 28, 29, 30, 31, 32, 33, 36, 37, 38, 39, 42, 44, 48, 52, 53
divergence, 6
duration, 12, 51

E

elastic deformation, 15
energy, 1, 2, 4, 7, 12, 13, 15, 16, 17, 18, 19, 44, 59, 60, 61, 62
ethylene, 1
evolution, 26, 28

G

gel, 9
grains, 3, 5, 21, 50, 57
graph, 47
growth, 3, 4, 19, 22, 43, 45, 52, 53, 54, 62

H

homogeneity, 24
hysteresis, 2, 5, 15, 17
hysteresis loop, 5, 15, 17

I

impact energy, 11, 13, 15, 18, 19
indices, 27, 32, 38
inequality, 3, 36, 47, 62
integration, 37
interaction, 52, 53, 63

K

kinetic equations, 6, 26, 52, 53, 54, 63
kinetics, 42

L

leaks, 51
linear dependence, 15, 19
linear function, 59
liquids, vii, 1

M

matrix, 37, 40, 46, 47, 48, 49
measurement, 3, 11, 12, 13, 15, 18, 60, 61
media, vii, 1, 2, 3, 4, 5, 9, 11, 13, 15, 18, 21, 23, 25, 31, 36, 38, 39, 43, 44, 45, 50, 51, 53, 54, 55, 57, 58, 59, 60, 62
mercury, 1
model, 5, 7, 26, 52, 56, 59, 61, 62
momentum, 51
motion, 29

N

nanometer, 1
nonoutflow, 2

O

operator, 42

order, 2, 29, 31, 35, 40, 46, 47, 54
ores, 24

P

penetrability, 56
percolation, vii, 1, 2, 3, 4, 5, 6, 12, 17, 18, 19, 22, 24, 25, 27, 30, 31, 33, 34, 36, 38, 52, 53, 57, 58, 61
percolation theory, 2, 24, 30, 52
phase transitions, 2
polymer composites, 2
porosity, 22, 23, 24
porous media, vii, 1, 2, 3, 4, 5, 6, 9, 11, 13, 14, 15, 17, 18, 19, 21, 23, 24, 25, 26, 28, 30, 31, 33, 35, 36, 38, 39, 40, 42, 43, 44, 45, 47, 48, 50, 51, 52, 53, 54, 57, 58, 59, 61, 62
pressure, vii, 1, 2, 3, 4, 5, 6, 9, 11, 12, 14, 15, 16, 17, 18, 19, 21, 22, 23, 24, 25, 26, 27, 29, 30, 31, 33, 35, 38, 39, 41, 43, 45, 48, 49, 50, 51, 52, 54, 55, 56, 58, 60, 61, 62
probability, 22, 23, 24, 29, 58
pulse, 50, 51

R

radius, 2, 3, 9, 18, 23, 24
range, 10, 13, 18, 50, 53
reason, 5, 24, 25, 31, 63
relaxation, 40, 53
resistance, 1
retardation, 6, 52
room temperature, 58

S

salt, 1, 11
segregation, 11
sensitivity, 3, 10
shape, 2
shock, 4
silica, 9
silicon, 1
silochromes, 1

Index

simulation, 51
SiO$_2$, 9
skeleton, 1, 9, 24
space, 6, 22, 24, 25, 44, 53, 58
specific surface, 9
spectrum, 40
steel, 9
strain, 50
strength, 50
surface energy, 1, 18, 24, 55
surface layer, 3

T

tau, 46
temperature, 18, 19, 22, 24
threshold, vii, 1, 2, 3, 4, 5, 6, 11, 18, 22, 24, 31, 33, 34, 36, 38, 39, 52, 53, 54, 57, 61, 62
time periods, 46
tin, 12, 13, 18

transformation, 15, 26, 32
transition, vii, 1, 2, 3, 4, 5, 12, 17, 19, 24, 25, 38, 39, 47, 57

U

uniform, 3, 25, 58

V

velocity, 2, 55
viscosity, 5, 6, 18, 19, 22, 45, 54, 58, 61

W

wetting, vii, 1, 3, 15, 26

Z

zeolites, 1